From Corporate
to Cariboo

From Corporate to Cariboo

MY FIRST FIVE YEARS RUNNING A RESORT

Leanne Sallenback

Sallenback Press

Sallenback Press

Printed and bound in Canada by Friesens Corporation

This book is dedicated to my parents.
Mom, thanks for being my biggest supporter on this journey.
And Dad, I hope you are enjoying the view from above.
I feel fortunate to have parents who sought adventure-filled
lives and inspired me to always follow the path less travelled.

Contents

FOREWORD   xi

CHAPTER 1
Chapter 1 of Life 2.0   1

CHAPTER 2
The Adventure is Worth a Shot   11

CHAPTER 3
#WaterPressureGate2019   15

CHAPTER 4
Groundhog Day   19

CHAPTER 5
Expect the Unexpected—Every Time   24

CHAPTER 6
The First Winter   29

CHAPTER 7
Shitter's Full   36

CHAPTER 8
Spilled Milk Duds   41

CHAPTER 9
Wildfire Season '21   49

CHAPTER 10

Lake Corralling & Bats   60

CHAPTER 11

A Campground Manhunt   68

CHAPTER 12

Robyn, Ryker & the Backstreet Boys   72

CHAPTER 13

The Year of Water & "The Incident"   80

CHAPTER 14

The Year of Water—Part 2   88

CHAPTER 15

Reflecting on Five Years   96

ACKNOWLEDGEMENTS   103

ABOUT THE AUTHOR   105

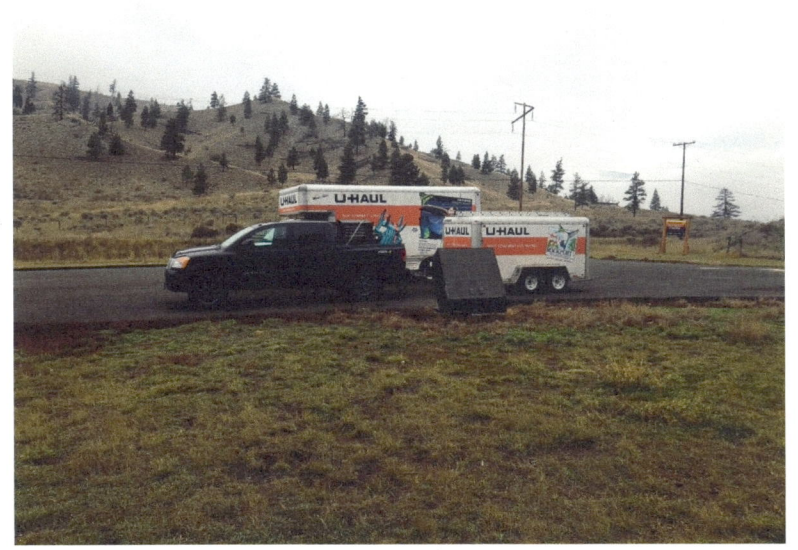

On the road from Corporate to Cariboo.
A rest stop in the Fraser Canyon.

Foreword

We did it.

In 2018, my husband Bear and I sold our house, quit our jobs, left the city and . . . bought a rustic resort?! That's right—we made the leap to the Cariboo region of British Columbia's Interior. Our next big adventure was taking over the historic Ponderosa Resort on Canim Lake, which had been in operation for over seventy years, and revamping it as our own, as South Point Resort.

They say the first five years in business are challenging—and they aren't lying. We faced wildfires, a global pandemic, travel advisories, resort closures, highway washouts, floods, major repairs, and various other setbacks. Spoiler alert: We are still married and our business is still operational, but along the way we faced sheer panic and uncertainty.

This story is meant to inspire people to break the mould, leave the mundane behind, and do something that scares them a little. If you are reading this book, I hope it inspires you to step outside the box and away from society's *prescribed* life. The unknown is a way more exciting place to dwell.

For me, the first step was leaving the comforts of pension security in corporate land and trading it all in for a life in the Cariboo. I was leaving behind something I was *supposed* to do in order to chase the unknown, which deep-down excited me. Still, I was sure that by following my authentic self, everything I ever wanted would present itself. And wow, did we ever manifest something incredible.

So, fasten your seatbelts, my journal and I are about to take you on a bumpy—fun, but bumpy!—ride. Enjoy.

*April 2019. The original owners, Wolfgang and Karin Martens
(left), and Bear and I (right) on completion day, standing in front
of the Ponderosa Resort—soon to become South Point Resort.*

Chapter 1 of Life 2.0

May 30, 2019

With my to-do list out of control, what better time to sit down and reflect on the last six months of my life? It's a blur, really, filled with moments of panic, excitement, stress, and joy. In November 2018, life as I knew it—on the Coast—ended. After working my way up to my dream corporate job, I quit. After spending three years renovating our dream Langley home, we sold it. And after thirty-four years living in the Lower Mainland, I left it. Yup . . . my life had turned into a country song.

At least my dog Ryker was alive to come with us.

This change wasn't a huge stretch for us. Bear and his family had been coming to Canim Lake for his entire life, and his parents (Val and Owen) retired up there. Bear had been bringing me up to Canim since we started dating in 2006. We tried to do a week or two each summer, and we were committed weekend warriors. We would race up to Canim after work on Friday afternoon and head back down late Sunday night. The travel was hard, but we loved coming to the lake.

Owen passed away in 2015, but Val still lived on the lake, and she factored into our decision to come up and help where we could. Bear's brother and sister-in-law (Ian and Karrie) also had a place on Canim, and his other two brothers (Warren and Jake) came up often. Canim was their family's happy place, and it had become mine.

I've always had an all-in personality, but I will admit, this was next-level. I went from boardroom presenter to resort owner; from worrying about my endorsements on LinkedIn and the success of my next strategic communications strategy to worrying about which covers fit which duvets, and which light bulbs I needed to convert to LED.

And you know what? I am so happy.

It's amazing how you get caught up in it all. Everyone says you should go to university, work hard, get the dream job, get married, buy a house, have some kids, and eventually retire, as if life has set bookends and you must follow the specified path between them. Gross.

I started all that, and it was the most unsatisfying thing ever. My outlook on life has always been a little different (a story for another time), so perhaps I was never destined to put in thirty years behind a desk. When Bear and I decided to move, I had been working for a large BC energy utility for over nine years. I was managing a marketing and communications group. It started to feel inauthentic, and I started to dread going to work, dealing with drama, and ultimately wasting my days doing something I simply was no longer passionate about. But, more importantly, I wanted to work for myself.

It was a hard decision, because I was surrounded by the best of friends and the best of colleagues, but you know when the universe is telling you to move on. Some people thought I was crazy to leave and give up the pension. The golden handcuffs just did not appeal to me. They were simply that—a trap. Others were happy for me, but I overheard some say I was making the biggest mistake of my life by giving up a pension and a six-figure salary. Meanwhile, I felt I was giving up a soul-crushing life sentence in a cubicle. It just wasn't for me.

At the time, I was living in a neighbourhood comparable to Pleasantville. Everyone had two-point-five kids, a dog, shiny cars, and immaculate yards. If I didn't clean up the leaves from my magnolia tree within a day or two of them dropping, the stroller brigade shunned me as they walked by. Our neighbour had the best grass on the street, and there was a *very* clear line down the property to mark the weed-infested, half-dead grass that belonged to us.

We were also dealing with some infertility issues that meant we wouldn't have kids ourselves. It was something we didn't talk about with people, but it was a huge deal for us. The stressors were out of control and mounting. We just couldn't try to keep up with it all.

Bear had just finished taking a year off to renovate our house. Literally everything was new, inside and out. The moment he mounted our twelve-foot, wall-to-wall, live-edge mantel in the living room, I knew the house didn't feel like home. We were trying to make a rustic cabin vibe in the middle of the most desired elementary school district in the city.

When I told Bear we were moving, he looked defeated but elated. He was going home to Canim.

A big part of the reason we wanted to move is that we had both lost our dads to cancer in recent years. My dad never got to see a day of his retirement, as he died about a month shy of his eligibility. And Bear's dad didn't get to enjoy his long enough. In a way, we thought we were doing this for them *and* for us. Because you never know when your time is up, so you might as well live for the moments that excite you.

Bear's dad once told me I was going to live at Canim one day. I replied, "Yeah right! I am so far from retirement." He smiled as if he knew something I didn't. I know who's laughing now.

The resort needed—and still needs—a lot of love. A lot of repairs and upgrades. We knew it was going to be a money pit . . . money we didn't have yet. We had to flip our house in Langley, use all our savings and some of my built-up pension, and we were still short. After writing detailed business plans and coming up with conservative forecasting for the resort, I submitted proposals to numerous banks. They all said no—we were way too high-risk. Oh my god. I thought this stage was going to give me an aneurism, and I was visibly aging by the day. How could these banks not believe in this business? So ridiculous!

Everyone was worried about me. Maybe it wasn't worth the stress, they said, or maybe it wasn't meant to be. How dare they? It was obviously meant to be! An unforeseen force was pushing and pulling me to the Cariboo. Plus, I had already quit my job. Can you imagine how embarrassing that would have been if I had to ask for my job back?

In the end, the sale worked out through private financing.

I dragged in my sister and brother-in-law (Corine and Matt) so we would have some more funds, convincing them they needed a cabin on the property as a recreation home. Luckily, that plan worked out.

We had no excess money, but we knew this was going to work out. I had woven the fabric to make the purchase a reality. However, I had no money for anything else—and there was a lot to do. Right on.

In the months leading up to our move, I shipped Bear away to work with his brother Ian for a couple of months to get some additional funds. He was so happy about that.

Before we left, we had one last night in our Langley house with our best friends who all lived around us—Sarah, Chris, Kelly, Rob, and their kids—and we did a champagne/orange juice toast to the next chapter. It felt like the last episode of *Friends;* we were all in the empty apartment, knowing everything was about to change. Everyone turned in their spare keys to our house, and we left them in a pile on the counter. These friends had been so influential in choosing that house and that location—and now we were leaving the security of it all behind. Ryker was already in the truck, waiting to go.

Our friend Chucky helped us move up the next day. We had a U-Haul, and the truck with another tow-behind trailer. It was go time. I started out driving the U-Haul, until I smoked a curb leaving the rental facility and made Chucky super uncomfortable, so he told me to pull over and he'd drive. Probably a wise move. The drive up the Canyon was surreal, as I knew I was driving to my new home. It was hard to believe what we'd just done, and I had no idea how any of it was going to go.

We came up here in a wintery February when it was -25°C and everything was frozen. We threw some stuff in the condos and cabins on the property and settled in for a few months with Val before moving into our log house.

The resort in the background after we unloaded everything
and moved in with Bear's mom for a couple of months.

It had only been a few months, but I had already seen positive changes in my life. I felt better; had way less stress in my life; got to be with Ryker all day, every day; and I got to take in beautiful Canim Lake whenever I wanted. I was fully aware that new challenges were coming my way. In the world of Google and Facebook reviews, I was expecting the good, bad, and ugly. But I'd take it in stride and make the best of this life. At the end of the day, we were running a resort. It's not rocket science, and people on vacation are generally happy. If the laundry machine doesn't work one day or the boat needs more gas, those were problems I was willing to solve.

I consider myself a first-phase Millennial. Some refer to us as Elder Millennials. I feel like I bridge the gap between Gen X and the other Millennials. My generation cares more about our vacations than our job titles, more about lifestyle than pay, and more about the next big adventure than anything else.

I have been fortunate. I've seen two of the seven Wonders of the World and I've dragged a backpack through four continents. I've climbed through the great pyramids in Egypt, sailed the Nile, surfed in Eastern Australia, hiked the Lares Trail in Peru, four-wheeled across the Salt Flats in Bolivia, visited a penguin colony in Southern Argentina, got lost in Japan, sailed the BC and Alaska coasts by small ship, and have a lifetime's worth of memories on the road. Travel has been a priority in my life. I was blessed with parents who travelled when they were younger and got into some crazy shenanigans, so I followed that path.

During my adventures around the world, I was always obsessed with hostels. I still remember all my favourites. From the Backpackers in La Paz, Bolivia, to Byron Bay, Australia, and the dozens in between. Maybe it was the vibe—rooftop decks, meeting new people—or just the non-responsibility of it all. I always took photos of the rooms, check-in desks, and layouts of their grounds. At South Point Resort, I have combined my favourite parts of all those hostels. My dream of being on vacation every day has finally happened. I get to be the person at the check-in desk and chat up all the travellers. How cool is that? Of course, now I'm also the one to clean the bathrooms, which is a bit of a switch up.

The previous owners, Wolfgang and Karin Martens, ran the property for forty-five years. It had been for sale for fifteen years, and it was time for new blood and revitalization. They had made a life here, and we were going to do so as well. Bear was good friends with their son, Rob, and we knew their daughter, Jessica, from visiting over the years. Bear's parents used to watch the property when the Martens were away, and vice versa. We already had a great connection.

I still remember the day I decided what we were going to do. I kayaked over from Val's property, walked up on the deck, and told Karin and Jessica that I was going to put in an offer. I think they thought I was joking, but I assured them that I was serious.

Jessica said, "Are you sure?! Have you thought about this?"

Three days later, I was back up for a showing, and the offer went in.

We lived with Val for the first few months, giving Wolfgang and Karin time to move out. We moved into our house on the property on April 11, 2019, and we opened on May first. No big deal. At the time, it seemed completely reasonable, but in reality, the timeline was so overwhelming.

Our family and friends showed up to help in so many ways. Sarah helped me go through the condo storage rooms and linens, and she raked a ton. She and Kelly really kept me calm. Rob painted the pink walls in the condos, helped Bear with docks, and did everything related to maintenance. My brother Clint and his wife Karen came up with my niece Evie to help with yard maintenance. Val made us meals and kept us fed. My mom, Corine, Matt—everyone pitched in. Bear's cousin Chris helped us a ton with excavator work and levelling things out. It really is a blur, but we were surrounded by help. My mind raced with everything that needed to be done, and I wanted to tackle it all, but I also wrestled with having zero money and being realistic.

I had spent the previous months moving all the existing bookings from a giant book (half in German) to my new online booking system on the cloud. I'll never forget sitting down with Karin at the table and going over the bookings for the year. She was stressed that I was entering it all on a laptop and not writing things down, and I was stressed that she wrote it all in this giant book that looked like it belonged in Hogwarts. I asked her if I could keep it afterward—it's an ancient artifact I'd love to display one day.

The Martens were great about showing us all the nuances of the resort. Bear and I decided to run it 'as is' in year one to learn, work the kinks out, and determine the investment priorities for year two. There were about fifty sets of keys, all colour-coded, and I filmed Wolfgang showing Bear the water lines and septic tanks. Wolfgang asked if I was going to take notes at all (confused we weren't), and

I said I was just going to film everything on my phone. He looked extremely nervous. It was a hilarious example of our age gap. When we put up the new South Point sign in the driveway, Wolfgang asked what the icons on the sign stood for—and I responded, "Those are social media icons for Instagram and Facebook."

We had some good laughs, and the turnover was smooth.

Things I've Learned So Far

1. You can use up your monthly Wi-Fi package in three days. Yup. There's nothing a little Facebook Live, social media, and watching one episode of *Project Blue Book* on Netflix can't do. So, I've had to learn how to scale back my time online to accommodate satellite Wi-Fi.

2. I was never one of those girls who wanted a white picket fence. Well, I got one anyway. It's about 400 metres long . . . and you have to rake under it every year. Fact: It takes four people two full days to rake and carry the leaves away.

3. Spring break in the Cariboo is just a fancy word for mud. If you have a dog, you know what this means.

4. If you go for a brisk walk on a -25°C day, your eyelashes can freeze.

5. It's a thirty-five-minute drive to town. I was used to driving over 20,000 kilometres a year on the Coast, but now I have "town days." I embrace them. I do all my errands in one go. It's like a race, really, to finish your list. I leave town feeling accomplished each time.

6. "Cariboo time" is a real time.

7. Asking for a "cord" of wood is common.

8. I have a landline phone for the first time since the house I grew up in. It's so weird. For the first five days, I didn't have call display, and it was the scariest thing ever. I froze every time the

phone rang, thinking people were going to prank me. I have call display now. Life is good again.

9. Wooden drying racks for your clothes are like gold up here.

10. I thought a lot of people were smoking fish up here, but it turns out there are outdoor wood-fired heaters for your home. Who knew?

11. There are amazing people in this town. I have already met some stellar folks who I hope to keep in my life forever. You know who you are. You've helped me with so many things during the transition up here, and I appreciate that you've accepted me into this community.

Standing at the bottom of Canim Falls.

CHAPTER 2

The Adventure is Worth a Shot

May 30, 2019

For the last month, I've been living like a lyric out of an upbeat country music song. I've found myself in the heart of rural life and a cliché like no other. I've caught myself bringing Bear sweet tea while he's on the tractor, hanging out laundry on the line every day, fishing, getting up with the sunshine . . . and loving every minute of it.

Last summer, as we were putting in the offer for this place, a song hit the airwaves that became my theme song: "Simple" by Florida Georgia Line. Man, it hit home. The whole idea was that life shouldn't be over-complicated, and we need to go back to just being humans—not robots. The song was everything I wanted our lives to become and, well, we manifested it.

At the time, the whole idea of this big adventure seemed like it was worth a shot.

This past May long weekend was full-blown business. We had a family reunion (amazing people!), and our own friends and family came up here to help us out. Our best friends bought an empty lot beside us. It felt like we were growing this little life into something special.

We even hosted our first annual yard sale and silent auction. It all went well, but for some reason, I started to get nervous afterward. I did have a fleeting moment where I asked myself if I could do this for the next thirty years. I wasn't sure, but then Carol and Al drove in on Tuesday and everything changed.

They stayed in Cabin 1, and we got to know them quite well. Over late-night bonfire chats, they showed us how to play their yard game, "Washers," which Al had built himself ten years ago. They were the most down-to-earth people, and we clicked instantly. They basically reassured me that this was the best decision we'd ever made and gave us all the confidence in the world that our resort was going to be a huge success. Not that I didn't think so, but man—it's been a big change, and I was still nervous about paying the mortgage.

When I went in to clean the cabin after they left, there was the Washers game we had spent three days playing, left for us. Unreal. That was the moment I knew this whole adventure was worth the shot (pun intended). Thanks, guys.

I decided to learn how to live with these types of moments here in the Cariboo. You know, the ones that could make a three-minute country music video highlight reel. Forget the nonsense that dwells in the B-roll and find those gem moments. I'm looking forward to all the future Carols and Als we'll meet throughout the years. We get to live on the lake and share our lives with folks from all over the world. My key is going to be: Always be creative, but keep it "Simple." Don't overcomplicate life.

Recent Learnings

1. Folding the fitted sheet. Am I the only one who didn't know there was a correct way to fold these sheets!? Omg. I will admit, this was life-altering. Laundering bedding has never been one of my perfected skill sets. Thank you, YouTube, for clearly explaining, step-by-step, how to achieve this goal. Martha Stewart, watch out.

2. Don't hang laundry out on the line when the sap (or whatever that is) from the cottonwood trees is falling. You will simply have to redo your laundry. This is a true test of your patience.

3.   Don't laugh at mosquito jackets. Whoever invented those should
     win a Nobel Prize.

4.   If you clean off someone's patio and a bird takes a massive shit
     on it immediately afterward, remain calm. It's easier to clean
     when it's fresh.

5.   Lake trout . . . have sharp teeth.

Bear and I relaxing on the water after the stressful events of the day.

CHAPTER 3

#WaterPressureGate2019

July 17, 2019

My thirty-fifth birthday turned into somewhat of a campsite gong show. Three toilets were plugged, and a massive water pressure issue presented itself as a unique birthday gift that we now refer to as #WaterPressureGate2019. The intense thunderstorm that rolled through that morning should have been a sign from the universe that things were about to get real. Because they sure did.

A large group arrived for a weeklong stay—and right as they all pulled in with their RVs and unloaded their gear, we magically had no water pressure.

Those are the moments when you realize how quickly you need to fix things. I pictured the zombie apocalypse on the horizon, with all the guests walking towards me in need of water instead of brains. If I'd been on the bridge of the *Enterprise*, I would have had the shields up and red alert activated.

Having to tell the whole campground the day they arrived "We have to shut off the water completely because we are not sure what the problem is" is not ideal. Running around and looking for underground water leaks in the campground when it was pouring rain was not an easy task. Everything was already wet, and I didn't really know what I was looking for. After some deer-in-the-headlights looks (from me) and some quick troubleshooting, Bear solved the problem.

In the end, the guests were great about it—and for that I was super thankful. The whole event was like built-in marriage counselling. If you can survive running a campground/resort together, as a couple, I'm ninety percent sure you could successfully take on all the Avengers at the same time and win. Bear and I started learning this early, and we knew our team skills would be an asset during future inevitable disasters.

But back to my birthday.

Our friends Braden and Sam were up with their kids. They, plus Kelly and Rob and their family, were all waiting by the campfire with cake and goodies for me. Everyone had strung up lights on our deck, the kids all made cards, and the bonfire was roaring. My birthday turned into an epic night.

When my friends and family asked, "Are you okay?" I always responded with "yes"—even when I wasn't. In the first few months of running our resort, I learned how to balance family, friends, and visitors with guests and the shitstorms that arose. Really, it's all about perspective and ensuring you spend quality time with people. The kids are a staple around the resort now, and it is super fun watching them enjoy the lake and all it has to offer.

To top off all the excitement, our first July was an absolute wash-out with rain. Not just any rain—it was like *Jurassic Park* monsoon rain. We're grateful we haven't had forest fires, but man—it's been muddy . . . and wet.

We knew the campground needed gravel and the sites needed some work, but we heard it was supposed to be the hottest summer on record, so we thought we'd budget for gravel the following year. This was a mistake. With the rain, all our roads and slips turned into mud slicks. We probably could have hosted the first annual Cariboo campground mud-wrestling event, but alas . . . we didn't pull that trigger.

This year, we hired our first employee ever, Sarah, to help me with housekeeping and the office. I don't know why, in the first year, I decided to have office hours from 9 AM to 9 PM *and* do all the housekeeping while having only one part-time employee. I am never doing that again.

Overall, I think we've handled ourselves well. There have definitely been some sleepless nights and moments of pure panic, but we're still in business! The forecast looks like summer is on its way, and let me tell ya, I can't wait.

Here's to the remainder of the summer being full of water pressure, sunshine, and my new Yeti stuffed with cool beverages.

Recent Learnings

1. A guest caught a fifteen-pound lake trout that was larger than most of the children who came to the cleaning table to check it out.

2. Our property is full of four-leaf clovers that the kids bring me all the time for good luck.

3. I evicted my first guest! A small bat flew into my house and spent two nights in the logs on the vaulted ceilings. I finally outwitted him and got him out (alive). Note: Bats don't like the sound of a bug zapper. If I kill mosquitos in my home with the zapper, the bats will start flying. That was a fun afternoon.

4. When looking for dew worms, always use a flashlight and not the light on your iPhone. It casts too much light to catch them.

Adapting to my new life in the Cariboo after a crazy workday.

CHAPTER 4

Groundhog Day

September 2, 2019

"Okay, campers, rise and shine! And don't forget your booties, because it's coooold outside!"

That is a line out of the 1993 classic *Groundhog Day*, which for the record, represents the last 120 days of my life. Today is September second, and it's the last day of our 2019 summer season. It has been an absolute blur because we worked every day. But man . . . what an amazing year.

This time last year, we were signing the papers for this place, trying to sell our house in Langley, and figuring out when and how to tell my employer I was leaving. Today I sit here after the craziest year of my life and am super happy about our decisions. But for the sake of disclosure—and just for some insight—here is a high-level summary of how my days generally played out for the last four months.

Get up at 6 AM, have coffee, clean the bathrooms and showers, start checking folks out, turnovers . . . more laundry, social media, more cleaning, lunch, check-ins, more bedding requests, cleaning, start thinking about dinner, dinner, clean bathrooms and showers, final rounds with guests, social media, check quiet times around the campsite, catch up on emails, prep the laundry for the next day, and go to bed at 11 PM. Social media. Repeat.

And that was just *my* day. Bear's days were more on the maintenance, lawn-care, and garbage collection side. We basically saw each other at mealtimes and in the evenings. Or when we had to troubleshoot some unique scenarios.

My daily life could have rivalled Groundhog Day, for sure. But in the middle of my total auto-pilot mode, I had great conversations with guests and a lot of laughs. I also tried to get out for a couple of minutes here and there to sit by the lake, get on the boat, or just bask in the sun on my deck. Mostly, I just hugged Ryker.

We had the most epic August long weekend at the resort. We hired a local band called Dutch Courage to rock it out in the Tack Shed. They played some amazing classic rock and country tunes and kept the dance floor packed. Orville's Septic showed up during the party with a new, freshly painted cedar outhouse. Ron and his team unloaded it. He told me there were too many people at the party for our septic system, so he thought he'd better bring it. Where else would that happen? So amazing. I was really starting to love our community.

Our friends Dave and Alex hosted a Quiet Events party in our front yard afterward. Anyone who wanted to do so could come over, put on their headphones, and listen to whatever channel of music they wanted. It looked so cool with all the headphones lit up in the dark around the campfire, everyone jamming out to their own tunes.

Dave and Alex own Reveal Events Group, and we started brainstorming for future events at the resort.

That weekend was the first time I truly saw the northern lights in action. I went outside at about 1 AM to let the dog out, and when I looked up, I saw every shade of green dancing across the sky. It was a moment that will stick with me, because it was the perfect end to the first big weekend of our new life.

This summer I met the Carter family, who had been coming to the resort for twenty years: Linda, her mom Dianne, her brother Mark, and her two daughters Kacie and Kisha. They stayed in Cabin 1 for two and a half weeks every summer. They used to own a property on the lake, but after it sold, they started coming to the resort.

I remember talking to Dianne when we hadn't even moved into the house yet. I was in Val's basement, trying to take deposits, and

Dianne asked me to call her. I swear it was like an interview of her getting to know me. She assured me that this was their cabin, and they would be coming every year.

When I met Dianne, we had an immediate connection. We were both obsessed with sky-watching and talking about aliens. I look forward to many more years of this. So far, there have been two campfires where we all saw something in the sky that we couldn't explain. One was in the daylight, and one was at night. Those are the times we'll never forget.

Running a resort is a true test of the mind, body, and soul. You must be on 24/7—and sometimes give in to some ridiculous requests, mostly because you don't have the energy to fight them.

I don't know why, but shit always hits the fan in the moments of pure exhaustion. There have been a couple of doorbell rings in the middle of the night. Looking back, if I were a selfie person, those would have been stellar moments to get a shot of me. Two in the morning, wild hair, slightly flustered, and a fifty percent chance that I'm going into pure panic mode, depending on what they say when I open the door. My calm-under-pressure look is pretty dialed in, though, so even if I was panicking inside, I always pretended everything was totally under control.

Dealing with the public on the daily is mostly great, but sometimes tough. I had my first one-star rating on TripAdvisor, which was mildly devastating—but I look at all the other amazing reviews we're getting, and realize not everyone is going to like your place. There isn't anything you can do about it. And that's okay. So many people have been so supportive and complimentary—and for that, I thank you all.

Recent Learnings

1. I learned what an o-ring for a propane tank is ... and the only two dealers in 100 Mile House that sell them.

2. Mushrooms can grow larger than a small child.

3. You can't plant pumpkin seeds in August and expect them to be ready for October. That was super disappointing. Like over-the-top upsetting.

4. There exists a hair dye that will actually stain a whole shower stall pink. That can't be good for the body. Or my patience.

5. A family of two can go through two toilet paper rolls in one night. #impressive.

6. If you want to defrost a fridge freezer, just unplug it by accident. The mess it leaves is also super fun to clean up when you're in a hurry.

7. Bleach counter cleaner can turn your favourite jeans into something out of the 1960s. Why wouldn't THAT warning label be larger?

8. There are some amazing ghost stories around these parts (but I'll only share these at campfires!).

9. Pack rats, for whatever reason, look you directly in the eye for long periods of time when you're trapped in the same corner with them.

10. Don't have arguments with your husband on walkie-talkies. Apparently, a lot of other random people in the area can be on that same channel and don't care that you forgot to defrost the chicken.

A typical fall day at the resort.
Photo by Robert Brunet Aerials.

CHAPTER 5

Expect the Unexpected—
Every Time

November 17, 2019

Yesterday, I bought an ice-fishing tent. At the checkout, I realized my life would never be the same. Not only did I have that tent, but I also had some saw blades, ice grippers, and a compressor. My shopping patterns have definitely changed since I moved up here. I now race to the store when ice-fishing gear is on sale and when there are good deals on blades so I can make Christmas tree ornaments out of real birch trees we've fallen on the property. What is happening?!

Remember the movie *Grumpy Old Men*, where they sit around with their ice-fishing huts and bitch about everything? I always assumed that was what it was all about. Bear assured me it wasn't . . . and after speaking with some locals, I have realized it sounds really fun. They crank the country tunes, cook up some meals, and enjoy the sauna-esque day in their huts. What?! I can't wait to get out there, and yes, I will document this entire experience.

Everyone asks what I'm doing these days, so here's a glimpse into it. The fall has been really busy thus far. We've been landscaping areas for wedding ceremonies, raking up a storm, planning our 2020 season, renovating, burning branches, attending tourism events—oh, and training a new, feisty little cattle dog puppy, Roxy.

The thing that has most amazed me is that whenever we plan an event or something, there seems to be an aura of drama right before. Remember the summer concert? For those that were here, the God of Thunder opened right as we kicked off.

Then, we planned a weekend-long Halloween event, which was so awesome—but the first day of the event was one of those days I ran around like an absolutely crazy person, not knowing what to do.

I meticulously put up all my Halloween decorations a couple of days beforehand. You know, spreading spider webs in the trees, raking all the leaves into a cemetery location with all the gravestones, that kind of thing. I had things on the trees, hanging from the trees, and decorations everywhere. Well, this time, the God of Wind came out to play, and I literally lost everything. There was a moment when I turned to Ryker and said, "We're not in Kansas anymore, buddy."

By the way, I apologize if you were the lucky neighbour who had spider webs land everywhere on your property.

I tried my best to redecorate the day before, but it was still so windy that we held off until the bitter end. Corine came up to help. The morning everyone was supposed to check in, the power went out. I remained calm for the first hour, but by hour two (as I'm sure Corine could attest), I became a raging lunatic (internally, of course). I was doing turnovers in between. Making beds and trying to clean in the dark with a lantern is super fun and not at all scary when you've been researching ghost tales for a week. The power was off from 9 AM until about 4 PM, and then luckily BC Hydro had things up and running.

The guests who arrived early were really good about it, and it became part of the Halloween experience as they got lanterns with their keys to now-freezing condo rooms. I quickly vacuumed all the units when the power came back on and before people checked in, so I was literally in a full-on sweat when everyone was arriving. I covered my crazy hair with an equally crazy witch's hat.

The first guest to arrive was a gal named Tanya. She instantly told me she was a Gemini, there was a crescent moon, and it all meant something powerful. We hit it off instantly, and what I didn't realize in that moment was that Tanya would become our employee in 2020 and stay with us for the next several years.

Up here, what it really boils down to is you can't predict anything. Ever. It's all about adaptability. Things are always going to go a little off the rails—but at the end of the day, things always work out somehow. And everything turns into a hilarious story.

I feel like I'm starting to figure it out a little more. As someone who was always overly punctual, I'm learning to embrace Cariboo time . . . somewhat. I at least understand it. I feel myself slowly letting go of my Lower Mainland scheduled approach to things.

Canim is a beautiful place to live, and we are starting to make new friends and establish an awesome life up here. Even though we live in a remote area, I never feel like it's too remote. There is always something to do. If anything, I have developed even more ideas for what I want to accomplish over the next several years. If I could shut my brain down I would, but alas, the wheels spin. Bear hates that I sleep with a notebook on my bedside table because I usually wake up every night with an idea that seems over-the-top ridiculous to him.

As of February, I will have been living up here for a year, so I can't wait to reflect on that. This past September we had a locals night on the beach, and all the neighbours came to celebrate the end of the season with us. We had some of Granny Graces pies (from Canim Lake General Store), some appetizers, and some great campfire chats. My biggest takeaway is the community was cheering for us, and we were going to keep going with this South Point dream of ours.

Recent Learnings

1. Bear bangers are a lifesaver. Literally. Especially when you take your dog out to pee at 10 PM, turn around, and find a bear standing six feet behind you. Being armed with the right tools is never a bad thing.

2. Moose are freakishly tall when you're right beside them. Especially when you're in the truck, passing them on the road, and when you slow down, they just watch you like they know

something you don't. I was *still* looking up at the gal as I passed
her ... in the truck!

3. Wear thick gloves when you rake. Those lame thin Costco-type
gardening gloves will give you blisters like you've never seen.

4. When you bring in the docks with your husband and he's in the
boat and you're onshore, make sure you go over the hand signals
and terminology of everything before you start. Otherwise, you
will just seem like you're having a huge domestic—and sound
carries on the lake.

5. Pumpkins can't freeze or they will rot. Good lesson when you're
trying to plan an event and you have thirty-plus pumpkins outside in minus temperatures.

6. When you take the ATV with the trailer full of leaves across the
street, unhook the tie-downs to take it off, and dump the trailer,
make sure you don't drop the tie-downs in the leaf pile. For the
record, it will be thirty-four minutes before you find them.

Bear, Ryker, Roxy, and I getting an ice-fishing tent ready for a guest.

CHAPTER 6

The First Winter

April 16, 2020

What better time for writing a journal entry than now? The world is all isolating in their homes and it's a great time to reflect. My last post was about Halloween. Since then, we have had an eventful and amazing first winter season full of hilarious stories.

In the fall, our business picked up two awards. The South Cariboo Chamber of Commerce voted us "Best New Business 2019," and the Cariboo Chilcotin Coast Tourism Association gave us an award for "Best Digital Marketing 2019." We were over the moon with this recognition—and it was motivation to keep going. We hoped people saw the work we were putting in, but you just don't know until they tell you.

New Year's Eve was one for the books. Tanya came back as a guest and planted the seed that she would be looking for seasonal work if we had anything. I don't know if she saw the dark circles under my eyes or sensed my three ulcers, but I quickly said, "Yes, let's keep in touch." She was off to Europe but messaged me every-so-often about spring work.

In addition to Tanya, we had a full house for New Year's Eve . . . and a snowstorm FOR THE AGES that resulted in a twenty-four-hour power outage for our guests. It was like that heavy, coastal snow. Luckily, the temperatures were mild, but we still had our challenges. When the power went out, we thought, *Okay, let's hope it's back on by dinner.* When it wasn't on by dinner, we realized we had to fire up the massive generator in the condo building for the first time. For the record, this thing looks like it's from the prehistoric ages.

I got out my handwritten notes from a year ago and my videos of Wolfgang explaining how to start up the monster. With help from our friend Chucky, Bear got it fired up quick. It was as loud as a small jet plane, so that provided good ambiance for our guests. I had planned to host an epic New Year's Eve party that night and had over twenty-five additional people coming to join our thirty-plus resort guests. We had fireworks ready, and I had done up a party room with lights, décor, and selfie stations. I was ready to ring in the new decade in style.

And I'll be ready to do it all again next year . . . because we didn't get the chance. The fireworks were great, and even though some guests had to leave early, it was a memorable experience. I feel like we are now prepared for a lot of weather-related scenarios, and we have since purchased some smaller generators for the property to ensure we can at least run coffee makers. That was direct feedback from our New Year's Eve guest, Ross.

Family Day weekend in February brought our next big group. It was a great weekend of snowmobiles, ice fishing, skating, and tons of lake activities. February is the perfect month to experience a Canim winter. The fish are biting, the lake is frozen, and there is so much to do. Even Canim Falls is spectacular at that time.

A repeat group of Williams Lakers taught me a lot about burbot fishing, including how to properly skin the fish—"Like pulling off a sock." I had fallen out on the ice the week before this event and smoked my head. I hit it so hard that guests came out of their ice-fishing huts and asked if I was okay. They seemed more concerned than I was, but I definitely stayed awake for the next twelve hours just in case.

We decided to stay open this winter to test our facilities and see what it was like having a resort open year-round. January and February did not disappoint. Bear and I bought our own ice-fishing tent with a heater and all the gear. So much fun. Who would have thought!?

*My first time up at Mica Mountain
snowmobiling with Bear.*

I will admit that it was a glamping version of ice fishing, as I could still access the resort Wi-Fi, listen to my tunes, and check my social feeds as I waited. Bear is more dedicated—and slightly appalled that I multitask in the tent. I mean, there's only so much staring at the rod you can do, and Bear and I don't enjoy the same music, so we can't even put on the tunes. I'm mostly into country music and he's a diehard death-metalhead. He also told me loud sounds can deter the fish . . . so *Jesus.* What are you supposed to do? I resorted to taking epic sunset shots and filling up my Amazon cart with things I knew I'd never actually buy. And going through resort ideas with him. It's like the ice-fishing tent became our office.

Canim Lake has freshwater lingcod, also known as burbot. It's a great-tasting fish, but ugly as all hell. The first time I reeled one in, I will never forget the feeling. It wasn't joy—it was pure terror. If you have a fear of snakes like I do, I recommend you don't reel in a burbot. I was in the confined space of the tent with the door zipper done up. As I reeled it in, I saw the face, I saw the head, and then I saw the glistening of the SKIN as it thrashed into the tent! I thought I'd caught an anaconda as it flew through the hole, and I just screamed and screamed. I dropped the rod, desperately unzipped the tent door, and flew myself out. SO GROSS.

Bear was trying not to laugh, but he was also embarrassed by the psychopath resort owner that had just put on this display in front of our guests. He dealt with it. I mean, yeah, I over-reacted, but the thing was freaking massive (fifteen pounds and three feet long!), and the skin looked like a snake's. Our friend Cole always managed to catch the biggest burbot, so whenever he was on the ice, it was a good luck charm for catching fish.

We also built a skating rink, had glow-in-the-dark bocce, and enjoyed some super-moon nights out on the ice with guests. We were busier than we thought we'd be, and we know now what we want to do to make next winter even better.

But now, with Covid-19 delaying our 2020 season and threatening the rest of it, we're taking the time to complete some upgrades we had planned and pray for additional approved lines of credit. Insurance costs are up for resort owners this year and there is a Covid-19 pandemic. I think taking on MORE debt as a small business for sure makes the most sense right now.

Recent Learnings

1. When you decide to make an outdoor ice-rink and don't have a Zamboni—make sure you make it a reasonable size (not actual rink size), otherwise you will get a lot of exercise each day maintaining that beast.

2. When you have eight condo units, a cabin, and a house to keep lit during a power outage—you need at least thirty candles and ten lanterns.

3. Ice grippers are cheap life insurance. I fell really hard twice this winter—so I learned quick that a helmet and ice grippers can be instrumental in winter resort ownership survival.

4. When it's -39°C, your hair can freeze and break off if it's not completely dry when you go outside. #terrifying.

5. Assembling new furniture with your husband . . . always fun. Why do they give screwdrivers names like Phillips or Robertson? Why can't they just say star or square? It would save a ton of time.

6. Also, measuring tapes. I learned that saying sixty-seven inches and two dashes past the centre is not an acceptable way to let your husband know the length of something. I also learned that you shouldn't write a page-long document of measurements down for someone if you don't really know how to properly read a tape.

7. No matter how much you rake the property in the fall, when the snow melts it's like a wee leprechaun played a trick on you and dropped just as many branches and needles for you to rake up again.

8. When you auger a hole in the ice in front of guests, do it with confidence and don't show you're out of breath after the first fourteen inches of ice. Power through to twenty inches.

9. Make sure that when you get your resort insurance, "unpreced-ented events" are covered in the fine print. We learned that we are not covered in case of a pandemic. I am now going to be put-ting forward recommendations for coverage for a Sasquatch invasion, hostile alien takeovers, lake monster interruption insurance, and mosquito swarms or "Mosquitnado." You just *never* know.

Bear and I setting up for some burbot fishing one night.

CHAPTER 7

Shitter's Full

August 1, 2020

It has been an absolutely crazy year as we've all been working our way through the ever-changing world of Covid. For us it meant closing for a couple of months and reopening with restrictions in place—and a bit of uncertainty about what would happen during our let's-make-a-living season.

We made the personal decision to close from mid-March until June first, which we were not expecting, but we took that time to get a lot of things done. Painting, new beds in the condos, fifteen or more dump trucks of new gravel, and lots of landscaping. We hustled, and it's paying off because a lot of people are noticing it. After cancelling several weddings and reunions, we started to worry. To make matters worse, the water was extremely high this year . . . so we got to experience the joy of sandbagging for the first time.

I will admit, my first year in the Cariboo has been pretty extreme. Mother Nature just did not let up. We had a late winter and a monsoon spring. One day in June, it rained like I have never seen rain. The entire day it just pelted. On that magical day, we were sitting in our living room feeling terrible for our guests when a light drip fell from the ceiling onto Bear's head. Yes, our roof was leaking. And about twenty-four minutes later, I had a knock at the door from the couple in Condo 6, advising me that their unit had *two* leaks.

The couple was really great and actually said, "Oh don't worry, if you hold a bucket above your head when you pee it's all good."

OH. MY. GOD.

Of course the leak was in their bathroom right above the toilet. Very few responses will make any difference in that situation, so I apologized, gave them a gazillion towels, and prepared to ask for quotes for new roofing. (And yes, we are getting a new roof on the condos.)

That same day, we had a minor flash flood on the creek. I was preparing to evacuate some of the RVers because I worried there might be a log jam upstream that could burst the banks. I woke up that morning at 5:30 AM to a text from Val across the bay: "WAKE UP, FLASH FLOOD—BEACH CHAIRS FLOATING AWAY."

Let me tell ya—I was up and in that lake, picking out chairs like no one's business. I came back inside at 6 AM looking like I just swam in from Alcatraz. By this point, these types of incidents don't phase me as much, so I just put on a pot of coffee and send the dogs down to wake up Bear as I prepare my speaking points for him.

Being a resort owner is not always easy. There are scary situations. Kind of like the one I was in when I decided to write this journal entry. I sat down at the computer at 2:25 AM, because I was up dealing with a plug in a septic line. At 11 PM, we were advised that the condo septic was not working. This is what nightmares are made of for resort owners. You have a full house of guests, and the tank is full.

Let's rewind to 10 PM that night . . . I'd had a crazy day, so I literally fell asleep to a YouTube guided meditation. The doorbell rang at 11 PM. I ran upstairs to a guest saying, "The toilet is backing up and so is the bathtub." Not exactly the end-game type of Zen I was looking for when I chose the meditation video. But what do you do?

So, Bear and I got out the drawings of the septic tanks we had from the previous owners and went to work. I called the local septic company (Orville's . . . aka The Lifesavers) and was amazed to actually get a human, Bobbi, answering the phone at 11:15 PM. They instantly sent out Harrison. Bear and Harrison spent some time

digging, only to discover that the tank's lid was a prehistoric artifact that weighed way more than two guys could lift. They slightly
dropped it and it almost landed in the tank.

I called my brother-in-law Ian (at 2:20 AM) who lives down the
way to bring the John Deere and help lift it out. That's right—if I had
time, I would write a country music song about this. I remember
cringing as I called the house because of the early morning hour, but
Karrie answered the phone with a "What's wrong?"

I said, "No one died. I need Ian and the tractor. It's the septic."

She calmly replied, "He'll be right down."

I should let you know that Bear has zero tolerance for smells of
septic or anything rotting. I tried to give him a face mask covered in
essential oils, but he assured me that did not stop the smell. Then he
made me knock on the doors to tell everyone that we had turned off
the water and they shouldn't flush their toilets . . . because he was
secretly worried about shit geysers.

Orville's showing up in the middle of the night to save the day.

At one point, as I was running across the driveway to get powdered laundry detergent to mark lines on the grass to indicate where the tank was, I glanced down at the boat launch and saw the northern lights dancing ever-so-faintly in the distance. I took a moment to stop and look up at the sky. Again . . . what are the chances! I'd been admiring it for about twenty seconds when Bear started loudly dry heaving in the background. Let me tell ya—I laughed pretty hard watching those northern lights. As I looked up at that sky, septic issues seemed low on the stress scale.

Oh, and don't forget that the ambiance for our guests was absolutely amazing, what with having a pump truck humming, a John Deere rolling, and a forty-two-year-old dry heaving. Everyone in the condo got the Shitter's Full free night special. We also learned that the low-flow toilets in the condos had contributed to the problem. Turbo flush toilets are in the budget for fall. We don't mess around.

Recent Learnings

1. When you're across the street loading firewood and a massive "twister-style" gust of wind starts swirling around and fir cones start pelting you, you jump in the truck as fast as you can and drive into an open space. Your life depends on it. Don't try to be Helen Hunt. Get out of the way.

2. When the cottonwoods drop their sap and those stickies are all over your dog—use coconut oil to get them off. Otherwise, it will take you fifty-six minutes.

3. If a guest eats Cheetos or Zesty Doritos in bed and wipes their hands on the bedding—just wash the sheets and put them in the donation bin for the SPCA. There is no way you're getting it out. No way.

4. Anything can happen at 11 PM, so don't go to bed feeling relaxed . . . ever.

5. Spruce beetles like flying directly into your hair.

6. There's actually a contraption/hook/thing on the vacuum to hold the cord so you don't run over it (who knew?).

7. When you wash paint brushes in the sink, make sure the paint is latex and not oil-based—or you will probably have to replace some sinks and spend some time calming down your husband.

CHAPTER 8

Spilled Milk Duds

December 26, 2020

It's been another few months of unbelievable happenings. Living in the Cariboo is by no means dull at the best of times, and in 2020, anything is possible. We had a great August and September—the weather was good, we managed to make it through the end of the busy season healthy, and guests were great about the Covid restrictions. I know they say the first couple of years in business are hard in general. Layering on a global pandemic has taken new-business-ownership to the next level of planning.

Reflecting on the last few months of the season, I find it hard to remember everything. We had our first wedding in September 2020. Travis and Hailey had a small wedding due to Covid restrictions, but they had a beauty of a night on the beach and partied in front of the condos. It was our first time meeting their family, and they were so much fun. I was so looking forward to partying with them that night, but as I was making dinner, I sliced my hand open. Clearly, I was distracted thinking about this wedding. I drove myself to the hospital with one hand after coating the other one in gauze. Bear stayed behind to oversee the happenings.

We all know he's not a social butterfly, so he likely stayed in our basement, praying no one would knock on the door. I was upset to miss the party, but I'm sure they didn't miss my presence. I was so happy for the two of them.

Joyce also came and stayed with us, bringing her son Ken and his partner Anna. Joyce used to run Minac Lodge on Canim Lake. We had a really fun night swapping back-in-the-day stories. We had faced many of the same problems, just with different levels of drama. She used to run a restaurant and it sounded like a very lively place. I absorbed as much as I could and recorded some stories on my phone to listen to later. I have grand visions of writing a book on Canim history one day, and Joyce gave me some epic material—and a box of newspaper clippings. She also drank me under the table, as we were doing gin and tonics. I will never forget that night around the campfire under the stars.

Ice fishing in front of the resort.

Thanksgiving was really great. We had guests in the condos and lots of RVs. Three families deep-fried turkeys on the beach, which we loved watching.

As I was talking to our friends Corrie and Darren, who were staying with us that weekend, we looked down the lake and saw whitecaps and dark clouds. There was no time to react. Out of nowhere, it was like the *Wizard of Oz* met *Armageddon.* The waves started rolling, rain pelted everyone, fir cones flew at people from the trees, and the wind took everything off the beach and threw it wherever it wanted. Guests' pop-up tents were lifted, chairs went flying, kids screamed. My first reaction was to run down the entire beach to see what I could save.

When I got to the condos, I found Tanya pulling resort beach chairs and one of the cabin's huge deck mats out of the lake. We just looked at each other and burst out laughing because the rain had soaked us and we had no idea what was happening. I left her there and ran down to the water, where I found John, one of our guests, trying to hold his boat so it didn't smash against the dock. Bear grabbed him and helped him off the dock. I tried to hold his boat, but the waves were so big. I made the call for everyone to get off the docks, telling them that we'd deal with any damages later. It was a very nerve-racking *am I going to have to pay for his boat?* moment for me.

John then alerted me that his adult son was out fishing in his tin boat. He pointed down the bay to a grey speck rocking wildly in the waves. Bear and I exchanged a pretty dialed in look that meant *this could be bad . . . red alert bad.*

Without thinking, Bear ran inside to get the pontoon boat keys in case we had to do a rescue. I got the binoculars and some lifejackets as he started the boat. We saw the grey speck getting bigger, so we knew John's son was coming in, thank goodness. It was incredible to watch this little tin boat surf into shore on the four-foot waves; John's son managed it with a notable amount of skill. The best part was that his dog was up at the front of the boat with the biggest grin, having the time of his life. When John's son got to shore, the wind died down, the rain let up, and I poured myself a strong drink.

Thanksgiving is definitely a time to be grateful.

I was really excited for winter and building our off-season market, so when the travel restrictions returned, my heart sank. Every time we tried to gain momentum or traction, something else was thrown at us. I thought maybe we could stay open by promoting to locals, but things started to happen that made me think otherwise.

Not seeing my friends from the Coast as much was also a struggle, and having to watch one of my best friends get married on Zoom was devastating—although I was so happy I could at least see it that way. Sarah and Chris tied the knot at the height of Covid with only their immediate family present. They looked so happy—and Bear and I got dressed up in our living room with champagne to witness the event over Zoom. They had come up several times to support us at the resort, and I can't wait to see them again as a married couple!

We have been slowly replacing items in the condos. Hopefully everyone enjoyed the comfy new beds this season. We also placed an order for brand-new sofas with pull-out beds for when rooms had additional guests. They were supposed to be delivered by mid-December, which would have been ideal for a December twentieth opening.

Bear was out snowmobiling one day, and I was bored—so I took a reciprocating saw to the wooden futon frames, threw them all into the condo parking lot and the dump trailer, and took them across the street. For the record, me doing this all by myself was epic. I should have had a Go-Pro on my head to film the whole thing.

As I added the last piece of wood to the raging burn pile I had created, my phone beeped with a new email. I opened it to find the local furniture store telling me my new sofas were delayed because of Covid and wouldn't arrive until February. I looked at the futon frames I had just massacred and lit on fire. I don't know why that email came in at that exact moment, but what can you do other than laugh? I sat across the street for an hour, watching it all burn down, before coming back inside to debate whether we should stay open without couches in the living rooms.

We want to make the most of every season, so we have a lot of toys to ensure we can get out in the mountains. Our friends Rob and Ang were up, and we thought we would all go snowmobiling up to

Yanks Peak, past Crooked Lake. Rob is the son of the previous owners of the resort, and he and Bear have always been close. We loaded the gear, the food, and the sleds and headed out there. We spent an awesome day exploring and had a great campfire. Rob and Bear wanted to try a trail, so we started our climb up this route. We ran into a trapper on a sled and asked him how the trail was. He replied, "It's all good. Just a couple of creek washouts of ice, but it's not bad."

We should have taken that as a bit of a red flag, but we didn't. We pressed on. Rob and Bear were leading, and I'll never forget when the trail just turned into solid ice. Rob bailed to one side in front of me, Bear to the other, and I just crept slow and steady past them as they screamed, "DON'T STOP!"

When we made it to the top of the hill, the views were like nothing I had ever seen. I suspect this was the feeling the boys always had while mountain riding, but it was rare for me to see these 360-degree views of the Cariboo hills on a beauty sunny day.

Me at the top of Yanks Peak after an epic ride up with Rob, Ang and Bear.

We rode around the top, which actually seemed a little sketchy since all the sides had cornices. The sun was starting to set, so we made the decision to head back down, which went amazingly well. We capped off the day with a bonfire and a meal.

Later that week, after a very early snowfall, I took Bear's dad's old truck to town to get groceries. The whole drive there and back, I was planning scenarios for a Covid winter season. Christmas, New Year's Eve, and Family Day were amazing events the year before, and I wanted to continue the momentum with the locals. I had already cancelled Halloween, which was a hard pill for me to swallow.

When I turned right on South Canim Lake Road, I was feeling good. I was listening to my country tunes and starting to think I could come up with safe winter plan.

As I turned the first corner, I hit a patch of black ice. I wasn't going fast, but I knew there was nothing I could do. I was ditch-bound for sure. As the old truck slowly slid into the ditch and onto its side, I just cringed. I was thinking of all the groceries in the back seat—the eggs!—and the potential scratches on the side. As I came to a halt, what I didn't account for was the open carton of Milk Duds in the cup holder. Those Milk Duds unleashed themselves inside the truck. I will never forget the echoing sound of Milk Duds peppering the windows. I took a deep breath, opened the door, and crawled 'up' from the truck and out of the ditch. As I stood on the side of the road, looking at the surreal scene in front of me, I realized we were going to be closed for the winter.

The entirety of 2020 felt like one big patch of black ice, with my life and business plans completely out of my control. I took this as a sign. Time to regroup. Shortly after the ditch incident, the provincial restrictions tightened up even more, which cemented our decision to remain closed. Again.

So, don't cry over spilled Milk Duds. It might be a clear sign from the universe to take a breath.

Recent Learnings

1. If you're clearing brush and happen to pick up a weird grass pile—don't analyze it, just throw it directly into the dump trailer. Otherwise, a small mouse might run up your arm, causing you to go into absolute panic, screaming and waving your arms in the air. And then you'll have a logging truck honk at you because he thinks you're waving.

2. If you run over a metal roasting stick with your ATV and the prong is stuck in your tire, pull it out quickly and put duct tape over it. That'll last until you can calmly tell your spouse what happened.

3. Ice can actually form in and around grass, causing an unassuming death trap at 2 AM when you take your dogs out to pee.

4. If you're burning endless piles of tree branches and sticks, make sure you park your new ATV far away. Otherwise, when the wind changes direction, you'll have small ash-shaped holes all over your seat.

5. Fireball and eggnog is probably the most underrated drink of all time. Thanks, Rebecca.

6. When you're stacking wood, make sure the end of the stack is like a big Jenga game. It prevents the whole thing from coming down.

7. If a deer runs out of the trees at Mach 10, passing you in a complete panic, don't wonder if the deer is okay . . . wonder what is chasing the deer.

The start of the South Canim Lake fire in
July 2021. Fire coming down towards the resort.

CHAPTER 9

Wildfire Season '21

November 20, 2021

Let's reflect on the most eventful time of this past year. It's still a total blur of anger, terror, and sheer awe at the events that took place. First, I dreamt of wildfires all winter. That's right. I would wake up having night terrors of wildfires. It was a weekly occurrence, so my intuition was telling me that we had better prepare. I had a very vivid dream in January and felt compelled to wake up at 2 AM and draw the fire I'd seen. It was in the distance, in the middle of the mountain down the lake, and I knew it was happening in the first week of July. I knew it was coming. I remembered the guests who were arriving that day in my dream, the airplanes, and the exact spot the fire would light up.

In the spring, I forced Bear to fireproof the property with me. I became *obsessed* with picking up every stick, branch . . . anything potentially flammable. He thought I was insane but went along with it. I cut all the branches on the trees from eye-level down, and I personally filled eleven dump trailers of branches, leaves, and anything that I thought looked dry. I got the fire pumps out early. I was ready. This was totally worth it, as the firefighters complimented my fireproofing later on.

A heat dome hit the province hard in the last week of June. The once-in-a-lifetime event was quite enjoyable for me. After not having had a vacation in a while due to Covid restrictions, I fully enjoyed basking in the over 30°C temperatures for the week and living in the lake. Of course, pure doom still loomed in the back of my head. I'll never forget checking the weather report on June 29 to see that a huge lightning storm was rolling in on June 30. My heart sank. This was it.

On June 29, I took my friend Jo and her husband Ranj out for a pontoon ride. I was trying to give them the tour and I could hardly speak. The air was electric already. Our friends Nicole and Richard rolled in the next morning, and everyone was just so happy to be together and enjoying the weather that I couldn't bear to express my fears. I had to go for a walk.

When I got to the top of the driveway, I looked down the road and couldn't believe my eyes. A large murder of crows was barrelling down the road *right* at me. I stood completely still just watching in awe. *What the hell is happening now?*

And then I saw it. The sign of the impending disaster was laid out right before me. One by one, those crows flew straight into the South Point Resort sign. It was as if they were warning me. Though dazed, they eventually flew away while I stood there pondering life. I calmly walked back down to the house and told Bear to double-check the gas in the fire pumps.

That evening, the longest and loudest thunder and lightning storm rolled through Canim Lake. After what seemed like hours of thunder rolling around us, hail pelting us, and hundreds of "lightning strike alerts in your area" on my app, we just waited. Everyone was on edge when they went to bed . The conditions were there, but we just prayed nothing would develop.

In the morning of July 1, we woke up to a fire just starting above my mother-in-law's house. Happy Canada Day. We called it in and, in no time, the local volunteer fire department had hauled ass up the hill and taken care of it. We all thought we had dodged a bullet. We were so thankful to those folks. We even dropped off ice cream and Gatorades for them because we were so relieved.

Well, that was just the beginning.

Within the next day, several other fires had started. One directly behind us, on the top of the mountain; one at Howard Lake; and one at Succour Lake—the exact spot down the lake where I'd drawn the fire after my dream. I went and grabbed the picture and held it up for Bear. He looked from the drawing, to me, to the fire on the mountain, and quickly went and got our extra fire hose from the garage.

We ended up telling the guests to leave shortly afterward. Some left on their own, as the fire raging behind us put everyone on edge. The BC Wildfire Service sent a representative who said, "Wildfires rarely move downslope."

I assured them that this was no ordinary mountain. A vortex of air regularly came off the mountain and moved toward the lake. Alas, this was not deemed a high priority. Over the next couple of days, the fire slowly grew from the size of a car to the size of a football field. Eventually, it consumed the mountainside and jumped from peak to peak.

The next month of my life passed like a dream . . . or nightmare. Some days, we sat listening to helicopters fill up with water right in front of South Point Resort every three minutes before dumping their loads on the ravine behind us and all the hot spots. I felt like I was in an episode of *M*A*S*H* because tensions were high and the plane and chopper traffic was frequent. Any time it got windy, we were out with the fire hoses, watering things down and exchanging uneasy looks.

At one point, I called Corine to bring up more pumps and food rations for us. We were scared to leave the property. We were just on alert at this time, but knew it could turn into an order and we'd be banished. Val and Tanya were taking care of Val's property, constantly watering things down. Bear's brother Warren came up to install pumps for her, as she was more in the danger zone than we were.

When Corine and Matt arrived, they were horrified to see the fire raging behind us while me and Bear were just casually having a drink outside in the smoke. It's amazing what becomes your new normal. They helped us set up some pumps, and they kept their cabin watered down. Most importantly, they fed us for five days while we aimlessly wandered the property, wondering what was going to happen with our lives.

Global News had already interviewed me, and I was told my ranting made Global National. The cameraman showed up at 6 am to ask me questions, and all I could think to say was "Is my hair, okay? Am I going to be on TV?" He reassured me that I looked okay, and we continued with the interview. The best part was that they edited what I was saying—ironically about the lack of information being told to residents—into a story about a poor resort owner who had "her life loaded on a boat." My friends kept calling, threatening to make "Leanne's on a boat" t-shirts after Global looped it on BC1 for days. Awesome.

My big debut on Global National talking about the wildfire situation.

I also gathered the community in the Tack Shed with some local representatives to get an update. In the back of my mind, I thought, "Well, at least the neighbours can see this new building before it burns down. I mean, our friend Colin just helped Bear turn it into amazing rustic building over the winter! We planned to start hosting weddings and community events. We even built a bar! I can't believe there's a chance we might lose it."

At this meeting, a local resident said, "Well, you know what this mountain is called, Leanne?" I shook my head. She continued, "It translates to Scary Mountain." How perfect. The deemed South Point Resort fire is raging behind us on Scary Mountain. I must loop back with her to find out why it's called that. Deep down, I'm hoping for a supernatural explanation.

Corine and Matt had to leave at this point, and Corine asked if there was anything I wanted to give her that couldn't fit in our truck. I had to think about it for a minute. I mean, we had the pontoon boat and the truck packed with our livelihoods—was there anything else I wanted to save if our place went down in a blaze of glory? Yes! I rushed downstairs and got my photo boxes of anything pre-2006 that hadn't yet made it to the cloud, and I opened the safe and handed her my signed Backstreet Boy drumsticks. God help me if I lost those.

It's amazing what you pack when you're on alert. And the longer you must pack, the more irrational your thought process gets. In the beginning, it's just food, clothes, and papers/passports. But as time goes on, you slowly cram more things into your truck. Like large thirty-packs of Hot Tamales and Whoppers. It got ridiculous.

My birthday was July 13. At this point, I was spending most of my days in an astral state, wondering what it might be like to spend a summer living on the pontoon boat if the resort disappeared. I even had a Pinterest board going for Pontoon Living. For the record, it would have been epic.

Our neighbours Rich and Heike were up to check on their place, and I think they could tell I wasn't in good shape. They offered to take me wakeboarding and give me a bit of a normal summer day, so I couldn't refuse. Even with the smoke, we had a splendid afternoon, and then our friends Corrie and Darren came by after and brought us dinner. Val made a cake. All seemed normal. I remember thinking *Okay, I can do this.*

About six hours later, at 4 AM on July 14 (to be exact), I got a call from the BC Wildfire Service. "Hey, Leanne. I have a favour to ask. Can you please go put eyes on the fire?"

Is no one watching the fire!?

I ran outside, past the condos, across the bridge, and looked up at the hill. The only thing I could compare it to was a scene from the movie *Volcano*. It looked like a massive volcano had spewed lava all over the mountain, and it was all streaming down to Val's.

I told the BC Wildfire Service, "Get here. *Now.*"

The fire behind us coming down the mountain toward our property.

Within a couple of hours, the evacuation order was in place for South Canim, and it became a scene from *Independence Day*. The cavalry arrived, we had an airshow for the ages, and it was all hands on deck. I stood up on the bank and wanted to give the inspirational President's Speech about not going quietly into the night, but no one seemed into it (look, I love that movie). They were just actioning, which was cool too. Most residents stayed because everyone had boats in the water, and no one would have needed rescuing. (Oh, and thanks to our Minister of Public Safety for publicly shaming us for trying to save our properties. That was super helpful.)

The structural protection teams ensured everyone had their buildings set up, and our resort became the landing zone for choppers. Fire guards had already gone in on the hill, and the teams had a good handle on controlling things. The structural protection team asked me to prioritize my buildings at the resort in terms of importance, and then their manager came back to double-check that I'd done it properly: "Leanne, is the first priority really the BAR? And then the FENCE? And then that OLD shed? And *THEN* your house and the cabins and condos?"

I quickly responded, "We spent a lot of time on the bar, and that fence is brand new. And in that old decrepit shed is all my new vinyl plank flooring for the condos, which I spent over $17,000 on, so do NOT let that burn. Bear and I will protect the other buildings." They obviously ignored me because we had more hoses on the condos and our log house. In that moment, that decision made perfect sense to me. I mean, that fence was a labour of love, but I was happy they were prioritizing our home.

One amazing thing that happened during all this was the huge sense of community that grew. Everyone who stayed behind banded together like we were the Goonies. Every night that the smoke wasn't too intense, we would have huge raft-ups with everyone's boats and watch the fire—sometimes with popcorn (or Hot Tamales). We floated and chatted for hours. We made a few new friends and got to know some old ones even better. And we watched friendships form around us. It was the ultimate team-building experience. Everyone made sure folks had groceries and we shared resources when we could. I'll never forget those nights on the lake with the Yellow Jacket pontoon front and centre—Gord, Bob, Natalie, Rob, Ken, Ron, Rhonda, Rick, and a few others—we have some core Canim memories from those raft-up nights.

The fire was visible on our hillside for thirty days.

A neighbour I hadn't met before, Barb, stopped by to give us some fresh strawberries from her garden. She said, "I am so sorry this is happening." That kind and random gesture made me break down for the first time. I could hardly talk, so through the tears I thanked her.

Residents watching the fires on the hillside.
Rafting up together in our pontoon boats.

In early August, the orders came off and we started welcoming guests. Dayna's group was the first to arrive for their weeklong stay, and Pat had brought the boat! So, life was good. The Carters came instantly to their cabin, and the summer seemed to go on as normal. A huge thanks to all the pilots and firefighters on the ground and to the equipment operators for putting in the guards. You guys are rock-stars. Especially the guys who were flying the massive orange mosquito-looking chopper every week. The Canim winds did not present themselves during all of July. At the end of the day, we got lucky, and we all knew it.

We still lost a whole month of peak-season revenue, though. Everyone was super understanding (mostly). Some wanted their deposits back right away, and at the time I didn't even have money to refund them. I had already used my remaining pension money to refund guests during Covid, so we were running out of options. We had spent most of our money on renos over the winter, so we were banking on our sold-out summer to replenish the accounts. I filed my insurance claim immediately, only to find out that I hadn't read the small print. No matter how much you lost, they had already established a "maximum" to be paid out, which was less than a third of what we actually lost. Thanks, insurance. Always a pleasure working with you.

Really, 2021 was wild. As a business owner it was challenging, but we have adapted. We closed for the winter because of Covid travel restrictions. Spring brought the provincial messaging of "Don't go the Interior" because of Covid. Those bans were lifted mid-June. Then, on July first, the "Don't go to the Interior because of wildfires" messaging started. That lasted right through to the end of summer. Quickly afterward, Covid travel restrictions started again. Oh, and now the roads were washed out from the BC Storm of the Century. So, you couldn't get here even if you wanted to.

I'm hoping we can open for January and February, but if the aliens land, please note that your deposits are non-refundable. I saw a hilarious meme about BC basically being a game of *Jumanji*—and at this point I don't think it's far off. We will have to move to a model where people have to buy insurance for these types of events, as we actually can't afford to refund everyone when Mother Nature decides to unleash her wrath.

As I write this, all access to 100 Mile House is cut off from the Lower Mainland. People are panic-buying at grocery stores. I have shifted my thoughts to spring flooding on our creek after the fires ravished the hillside. I'm about to contact the Ministry and anyone who will listen about flood mitigation. I'm even looking at purchasing those massive inflatable flood systems to divert water from the condos if the creek decides to rage behind us. And if it does? Well, I might just buy branded inner-tubes and create a fun waterpark at South Point Resort.

We wrapped up the year with a great September and a super fun Halloween weekend and locals' night. We also celebrated the fiftieth birthdays of two good friends (Colin and Rebecca) in the Tack Shed this December. We had a winter party for the ages. Nory and I decorated the Tack Shed in full Christmas splendour. I put every decoration I had into this, and it looked fantastic! Christmas trees, garland all over the rafters, tables, window decorations—we went all out.

We had a raging bonfire, and I curated an epic playlist on Spotify. It was full of old country classics for Colin and random hard rock for Rebecca. I think I put a newer Aerosmith song on there that she publicly shamed me for, but otherwise I think I came out ahead. Most of the community came out to wish them happy birthday. I don't remember much from that night . . . but I do remember the Cariboo Colin shot at 3 AM that the remaining party members had to take. Luckily, Bear avoided this. He usually slithers away from gatherings without telling anyone. I have no idea what time he went to bed. Colin had worked on the Tack Shed and done a ton of work for us, so it was cool to host him and Rebecca for this milestone. I mean, how many times can you drink at the bar you helped build?

Overall, this community rallied and was super supportive this year. Now we have some upgrades ahead of us, and we will reopen for a couple of months this winter. The extreme weather events keep reminding us that there's always another great story just around the corner.

Recent Learnings

1. Don't sleep naked during an evacuation alert/order. Knocks on the door are frequent.
2. People don't know to keep their boats away from choppers or water bombers in a lake setting.
3. Not everyone is a communications specialist. So, Leanne should calm down. Stuff is getting done, they just aren't talking about it.
4. I love wake-surfing. How did I not try it before this year? Thanks, Pat.
5. When your phone constantly beeps because of "lightning strikes in your area" alerts, just turn off your notifications. Monitoring them is not worth the stress.
6. Minimal downslope growth as a general predictor of wildfire behaviour does not apply when wildfires are near a major body of water. Like a large lake.

CHAPTER 10

Lake Corralling & Bats

May 2, 2022

Yesterday we had a massive windstorm. This is not unusual, as we have these storms all the time on Canim. The unusual bit was the sequence of events surrounding this day.

I was making coffee in the afternoon, and then I walked out onto the deck to look around and enjoy the view. This was early spring, so the ice was just off, and people were thinking about getting docks and boats in, but it was early in the season and we still had lots of time. Our neighbours Rich and Heike are amazing people, and we help watch some of their stuff in the off-season as they are right next door. This year, Rich had just said, "Can you watch our dock with the rising water?"

I said, "Of course," and we checked on it every couple of days.

Well, I had to do a double-take. As I was standing on the deck, I saw their dock—with their slide on it—whip around the corner, caught in the windstorm, floating past us at high speed. I couldn't believe it. This was the dock I told them I was watching, and I was responsible for it! I immediately went into *Baywatch* mode and decided I needed to rescue the dock before it went down the lake and over the falls. In my mind, this would be catastrophic. I take my responsibilities seriously.

Keep in mind that none of our boats were in the water because we didn't have our docks in yet. Bear was out in the yard, working on a gate, and I probably sounded like a deranged banshee as I yelled, "THEIR DOCK IS FLOATING AWAY!"

Bob took this great photo of Natalie (left), me (middle),
and Bear (right) quadding out by Mahood Lake.

I threw on a lifejacket and grabbed the only paddleboard I had inflated (which was a small green kid one). Bear tied a rope to the back of the paddleboard, intending to use it to pull me in once I got to the dock. We quickly formed a plan, and it was go-time. This was my moment. I paddled through the chop toward the dock.

Amazingly, I did not fall in. The wind was insane, so the chop was crazy. I had to stay on my knees and paddle because there was no way I was standing through that. I reached the dock, absolutely exhausted from the ordeal. After I hauled myself and the board up onto their dock, I yelled to Bear, "OKAY, PULL ME IN!"

And he yelled back, "I DROPPED THE ROPE."

I could have killed him. The thoughts that went through my mind were purely revenge-oriented. He didn't want to get his feet wet, so he tried jumping the creek with the rope . . . and let it go.

He made no attempt to get in the lake to retrieve the rope. Just pure defeat. I mean, if it were me, I would have at least waded in!

The last words I sternly said to him were "You'd better do something!" before the wind pulled me out of earshot. I watched him *slowly* walk away and hoped, for his sake, he was strategizing.

I was shellshocked, floating down the lake with the dock in gale force winds, without any other form of rescue. Luckily, the wind was pulling me toward Val's, so I paddled the whole time with my one paddle, trying to keep the dock headed in that direction. I had visions of ending up all the way out in the big part of the lake just getting rocked by the waves.

I hoped someone would salvage the situation before the waterfalls became an issue.

Bear was still out of sight, so I thought he must be doing something to help. The Canim Lake Facebook group was always super active, so I thought *Someone is going to see me and post something, right?* Tanya was cleaning in the condos. She would see me for sure. Nope.

Then I saw our painter, Alex, in his bright orange shirt, painting my mother-in-law's house. I screamed at him against the wind. He was up a ladder and never even turned around. He always had his ear buds in. *And where is Val!?* It wasn't like her not to notice something out of place on the lake.

I landed at her neighbour Angus's docks and grabbed some of his buoys from his yard so the dock I was on didn't smash into his. I intended to save all the structures from the wind. I ran over to Val's in my bare feet to get Alex, and he helped me tie the ropes around the dock. We had a good laugh at the whole situation, and once everything was under control, he went back to painting.

I collapsed, totally exhausted, and waited.

Bear took a while. He got a boat from across the street with help from our neighbour, Ken, and got the motor on the boat. Our friend Willy dropped by, so he and Bear made their way over in the boat. I will never forget Willy going "Hey, miss. Do you need some help over here?" and then it was all just a good story. We managed to tow the dock back, but the wind was so bad we tied it up in front of our place to keep a closer eye on it.

It was a beautiful sunny day, so during the chaos I got a great tan. All's well that ends well. My feet were cold and completely numb, but the dock was good to go. I emailed the neighbours the story, and they were completely appalled at me for even attempting the rescue, but we all laughed at the ridiculousness.

The random things that can happen on the lake are so funny. For example, Alex had done a lot of painting for us for in our first few years and had become a staple around Canim Lake. Earlier last year, he was over for a visit. He and Bear were talking down by the beach when all the nearby dogs went wild. We looked out and spotted two moose swimming by, trying to get to shore, but the dogs were deterring them. The moose turned around and started swimming down the lake in the opposite direction. It was a windy day and they were swimming against the wind, and we didn't want them to get exhausted and drown. So Alex and Bear jumped into the tin boat and proceeded to corral the two moose back to the beach just down from our campground. It was quite the scene! The guys kept their distance, but they guided the moose to Val's property so they could head ashore. They got a standing ovation from our campground, and it was pretty cool to watch.

Bear was getting into the habit of corralling things. In the winter, we had a group of ice fisherman and they forgot to secure their tents to the ice. One February day, suddenly Canim came to life and just started blowing. Our neighbours had borrowed our tent, and their kids were fishing inside when it decided to launch. It caused some PTSD for those kids, I'm sure. That tent blew all the way down the ice—but, amazingly, just as far as the Sallenback property where Ian and Karrie happened to be visiting Val. They helped pack it up, and we brought it back.

One of our guests was on a trip to town while his tent was blowing right down the middle of the lake! It looked like it could have rolled and tumbled forever. So, as the good hosts we are, Bear jumped in the side-by-side and floored it down the lake. I had just started watching the series *Yellowstone*, so this was as close to Rip Wheeler as it gets. It took him a while, but he got ahead of the tent and then jumped out to grab it. Somehow, Colin was around for this, so he helped Bear fold it up and bring it back, and I think they secured it. We got all the tents back to their places, but some fishing supplies, like buckets and ice scoops, were lost to the lake gods during that storm.

I'm also up to date on my rabies shots, because a bat assaulted me in our house. We'd left the doors open most of the day while we were working on some things. Later that night, I had a shower and blowdried my hair before bed. When I blowdry my hair, it becomes a lion's mane, so it was super staticky and huge. I was being nice by not turning on the lights because Bear was sleeping. As I felt my way down the hallway toward the bedroom, all of a sudden, out of the laundry room, a bat flew into the side of my face, peeled itself away as if trying to stop itself, and landed in my static-filled hair! It was shrieking, I was screaming . . . the entire scene was bad.

But if you ever wanted to know the answer to "Can a bat get stuck in your hair?" the answer is YES.

Bear didn't even turn on a light, didn't ask if I was okay, just said, "What was that?"

I explained what had happened. He seemed unfazed, laughed, and went back to bed. I Googled to see if maybe I should be concerned that this creature and I just had a full-on encounter. I was so tired—and the bat had escaped my sight at this point—that I went to bed.

In the morning, I dialled 811. After I explained the situation, the nurse quickly said, "Yes—you need to go to the emergency room and ask for a rabies shot."

Ugh! There were a lot of shots that day, and then I had to go back every week for a bit. So now, when I'm ghost hunting (I'll get into that later . . .), they send me into all the sketch places first because I'm up to date on rabies and tetanus thanks to Cariboo living.

We had an amazing sled season with Bob and Natalie, riding trails and exploring new areas. It's mostly about the food, these rides, so we have learned how to up our game when travelling with them. As much as it is about the experience, it is also about the views, the gourmet smokies (with mustard), and good friends. I'll never forget the one ride we had coming back from Mahood Lake. The sun was setting on the trail, the moon was out, and we came across a few moose on the trail. Living up in the Cariboo has its perks when you have moments like this. Bob snapped this amazing photo of me, Bear, and Natalie on the trail, and this one lives strong in the memory bank.

Sunset shot on the sled trail. On our way home for the night.

Recent Learnings

1. On a sunny day, the metal in your ice tent pegs can warm up and cause the surrounding ice to melt. So, check them consistently.

2. When lake levels rise quickly, check the docks more often.

3. Feet warmers in your shoes are a game-changer. Buy them in bulk.

4. Invest in heated clothing.

5. If your husband decides to take you to the mountains for a "sled training day," make sure the groomer has been out at least once in the last couple of weeks. Also ensure you have a strong marriage, or it won't survive the day.

Aerial shot of the campground. Photo by Robert Brunet Aerials.

CHAPTER 11

A Campground Manhunt

July 2022

I am writing this after a pretty wild event. I woke up around 3 AM to the sound of screaming. I woke Bear to see if he could hear it and he said, "You'd better go check that out."

I deal with all the campground drama or issues. If it was left to him, we would quickly be out of business.

I started to get out of bed, and it seemed to be getting louder, so he got up with me. We rushed upstairs, opened our front door, and looked toward our shed. Our neighbour Cheryl from the other side of the lake was like, "HE'S IN YOUR GARAGE!"

I had no idea what she was talking about or why she was shrieking so loudly. I asked, "Why is Ross in there?" I was so confused and sleep deprived that this was the only question that came to mind.

Cheryl responded, "ONE OF THE GUYS THAT STOLE THE SIDE-BY-SIDE!"

Suddenly, her husband, Ross, came running down the road with a flashlight, shirtless, and breathless as he explained that someone had stolen an ATV and had been scoping out their yard. "We jumped in our truck and chased them all the way here!" he exclaimed. Apparently, the two thieves, with Ross hot on their heels, had crashed the side-by-side and were now on the loose—running through our campground on foot.

I just muttered "Oh my god" under my breath. Bear instantly grabbed his keys, jumped in the truck, and drove up and down the street in stealth-mode, looking. I immediately called 911, said, "You better get here as fast as you possibly can," and hung up the phone.

What I didn't realize had already happened is so hilarious now, but at the moment, I was shocked.

Fortunately, my friend Kristy, who had rented a trailer for the week and was staying in one of the back sites near the road, had a perfect vantage point for all the action. Apparently, she woke up to the loud sound of an ATV crashing into the far driveway of the resort. Kristy looked out her one tiny window in the back of the trailer to see the crash. Two guys got out of the side-by-side, a truck flew in driven by a guy with no shirt (Ross), and a full-on chase through the campsite commenced.

At the time, Kristy didn't know who was good and who was bad—and she was in the zone of the campground where the Wi-Fi didn't work. She was full-on panicking in her trailer, watching campers jump out of trailers in their underwear, brandishing frying pans. Her brother-in-law is a police officer, so, dressed in his camping PJs, he chased one of the theives through the site, past our house, and down the beach.

I had no idea any of this had happened. I think Kristy's brother-in-law might have got a brief tackle in, but I can't be sure.

The cops showed up very quickly; I have to give them credit for that speed. I was expecting police dogs and a SWAT team or something, but it was one guy from Quesnel who was covering a shift that night and had never been out this way. Now he was in for a couple of hours of a full-on manhunt for two guys. Kristy came out of her trailer when she saw me with the cops to tell me what had happened. We were all looking in buildings and sheds, figuring they had already taken off down the road. I looked in Corine's basement and texted her to see if she was up. She never responded, so I let her sleep and decided to tell her the story in the morning. It never occurred to me that the suspects could be upstairs holding her hostage, but I thought about that later.

I also had to tell the police officer not to fingerprint the side-by-side, because Bear's prints were all over it. He was the one who moved it off the road. I'm not sure how this story ended, but the thieves definitely got away. It turned into a hot topic at the resort the next day with everyone swapping stories. Corine couldn't believe she slept through it all, and neither could I!

In all honesty, what are the chances this chase would end in the campground at South Point Resort at 3 AM? The next morning, over coffee, Kristy and I laughed hysterically about her version of events— right before my official statement with the 100 Mile Police.

Recent Learnings

1. Know the RCMP's exact response time, and account for that in emergency planning.

2. Never underestimate campers. They are willing to jump into any situation to help. With frying pans.

3. Make sure all your flashlights have new batteries or you will watch your husband lose his mind during an emergency situation when none of them work.

4. Always have running shoes by the door. During a manhunt, Crocs are not the appropriate footwear.

Ryker and I boating on Canim Lake.

CHAPTER 12

Robyn, Ryker &
the Backstreet Boys

October 15, 2022

What a whirlwind year 2022 has been, full of major highs and big lows.

In May, we had our opening weekend. We called it the Decade Dance. We charged ten bucks a ticket, had a bartender, and advertised far and wide. You could show up for a great time, dressed as your favourite decade. We had a fairly good turnout. Chase and Wyatt were our bouncers; they even put bracelets on everyone who paid. The crimped hair and over-the-top makeup from the 80s were all the rage that night. The campfire was once again the gathering place for most people. I dressed up as a 90s kid, rocking my Backstreet Boys t-shirt and wild hair. Rebecca won best costume—I think because of the wild heels she managed to pull off and her backcombed 80s hair.

All that day, I had been messaging back and forth with a potential guest, Robyn. She was travelling solo and could only do one night on her way through 100 Mile House. Eventually I just messaged back: "If you want to come, your key is in Condo 7, the door is open, there is a Decade Party in a large barn-like building on the property that you're welcome to join. You'll hear the music."

At about 11 PM, this girl around our age showed up, and I knew it was Robyn. It was one of those "I knew you in a past life" moments, and we all just became instant friends. She loved the key-is-in-the-room mentality and hit the dance floor with us.

That night was also the first time I danced on my own bar, which was a proud moment for me. A little "Dust on the Bottle" by David Lee Murphy came on the speaker, and I basically hurled myself up there like it was 2003 at Roosters Country Cabaret. Before I knew it, there was a neon yellow flash and Tanya was up there, then Corrie, and then Robyn. It was this epic moment between the four of us that I will never forget. The season had officially kicked off!

This was also the night the rule "You can only dance on the bar if Leanne is dancing on it first" was implemented. This rule is strongly enforced.

Something that I haven't really mentioned is that, in addition to running the resort, for the past several years I had also been pitching a television series with my sister Corine and best friend Kelly. We're paranormal investigators (that's right), and our team's name is Beyond the Haunting Investigations. We had been doing investigations together for over five years at this point and thought *There are no women in the genre, so why don't we pitch this?* During Covid, we took advantage of all the film festivals being online, and in 2021, during the wildfire days of Scary Mountain at South Point Resort, I pitched this idea at the Banff Film Festival.

Earlier in the year, in partnership with funding from the Cariboo Chilcotin Tourism Association, we filmed a short "Gold Rush Trail" pilot, so we had something to show the folks at the Festival. They liked the idea, and the concept grew from there. In June 2022, the three of us set off with a full film crew and the backing of a Canadian network to film a two-part documentary, *Haunted Gold Rush*, retracing the steps of the iconic British Columbia Gold Rush Trail. It would air later in 2022 on the T+E Network in Canada. This was a major win for the three of us, and we hoped it was just the beginning. We had an awesome production team, Small Army Entertainment, and two amazing producers, Sean de Vries and Stephen Sawchuk. It was go time.

Now I had to tell Bear I would be leaving. This meant reassuring him that we had excellent staff at the resort to take my place. Tanya could operate the resort in her sleep, and Nory had joined the office team and was super capable. They had everything under control.

Tanya had just celebrated her fiftieth birthday by booking out the resort for her family and friends, so I was glad I was around for that. What a party!

A couple of days after Tanya's birthday, I left for twelve days of filming. We made it work at the resort, and I got to have the incredible experience of retracing the history of the Gold Rush Trail from Yale all the way to Barkerville in a fast-paced journey with an amazing crew. When I got back to the resort, it was full-on summer season, so without missing a beat, I had to put aside the ghost-hunting gig and plunge into the deep end of resort operations. At this point, I was compartmentalizing my life into sections to keep sane and focused on what needed to get done. Ghost hunting deactivated, high-season resort tasks engaged.

The next month was hard for us, as we lost our Ryker. He was the cattle dog we had for over ten years together, and he was essentially our child. Looking back, it was surreal. We had an amazing week with him, lots of beaching, lots of pontoon rides and enjoying the summer in all his favourite places on Canim Lake, all without knowing it was his last week.

One day, I was in Kamloops getting supplies. When I came back, I grabbed Ryker to go outside. Just him. I left Roxy inside and he and I went out and sat on the dock to watch the sunset. He brought his ball. He came up right beside me and put his head on my shoulder. This was something he hadn't really done since he was a puppy. I didn't bring my phone out with me, so I couldn't capture the moment, but it is something I will never forget. Looking back, I didn't realize it was his goodbye. After a while, I threw his ball in the water and he did one last dock dive. We called it a night.

The next morning, he was having major breathing and heart issues, and we lost him later that day. The vet let me lie in the hallway with him for over six hours as we tried everything to save him. The hardest decision I have ever made was to put him down so the suffering would end. I still break every time I think of that day. The last thing I told him was that I would search one thousand lifetimes and ten thousand planets to find him again. And that is an understatement. Love is the only thing that can transcend space and time, and I know it because of Ryker.

Remember Barb who dropped the strawberries off during the most intense wildfire day? Well, I think she has excellent intuition. The morning after Ryker passed, she showed up with a delivery for the second time ever. As she handed over the fresh Canim strawberries and opened her mouth to ask how I was doing, I absolutely lost it. Through the tears I said thank you. She had no idea what was happening, but she nodded silently. *This lady probably thinks I'm a basket case!*

I filled her in afterward and she said, "I hope you aren't starting to associate strawberries with trauma." I'm not. It's probably some cosmic connection between us that we're unaware of. They were damn good strawberries too.

August was a blur. I was still mourning Ryker's loss, but toward the end of the month I buzzed down to Vancouver so I could attend the Backstreet Boys DNA concert with one of my best friends, Jen. Yup, we are diehard fans. I had purchased front-row seats and backstage passes for us. We never miss their concerts. This was concert number eleven, and the third time meeting them. I had this master plan to get a photo with them and a poster I'd made that said, "*Haunted Gold Rush* Fans." It would go viral on our social media channels, and I knew my producers would love it.

I had the whole speech and everything ready as I approached them for my meet-and-greet photo. Just then, a security guard came up and said, "Ma'am, you can't take posters in."

I explained that it wasn't for the concert, it was a marketing thing, and they could have it after the photo. The security guard tried to rip it from my arms, and I was so furious. *Is this security guard seriously taking a poster away from a thirty-eight-year-old woman?*

Well, he did. I was so bamboozled that I didn't know what to do—and then it was my turn at the meet-and-greet.

What happened next was probably the most embarrassing moment of my life, but I can't be sure because I think my soul left my body. I believe I proceeded to tell my favourite band that security took my poster—and that I wanted them to come ghost hunting with me on my new TV series. Nick Carter (who I have been in love with since I was *twelve*) cringed, and I started disassociating in pure disgust at myself. Then Kevin Richardson grabbed my hand and asked the name of the show. I muttered *"Haunted Gold Rush"* under my breath.

He said, "Tag me." I'm sure he thought I was a complete whack job. I never tagged him. One day, I hope to repeat this story to him in person. I'm pretty sure I'm the only girl in history to ask the Backstreet Boys to come ghost hunting, so I hope they at least tell stories about that crazy chick in Vancouver, Canada. And one day I will tag you, Kevin.

The concert was amazing, but I had to get back the next morning as we had our biggest wedding to date happening that weekend at the resort. Erika and Craig were having a large wedding, and they trusted us to host them, so it was full-on go-time. All hands on deck. I ended up getting so sick that I was literally out of commission for the first day of the weekend. Tanya stepped up her game and took control of it all. I bounced back the next day, and everything turned out to be amazing. Great weather and great people.

We had another September wedding, and then we rounded out the year with the amazing annual Pooli Family fishing group. I was so grateful to have had a good financial year so we could put some money back into the resort.

In early October, I took the pontoon boat out on Canim, as the water levels were extremely low. We were with Corrie, Darren, Chase, and Wyatt, and we were hunting for the Canim Lake Sunken Islands. We stopped at Reynolds Resort, where Dan gave us some stellar directions (it sounded like a Goonies pirate map): "Look for the three dead trees, then go to the point, then look for the house across the bay. There is a floating bottle."

Totally reasonable for a lake that's thirty-six kilometres long and over 700 feet deep in places. But we found the amazing Sunken Islands. I jumped out and asked Bear to take a photo of me as I pushed the boat away from me.

For some reason, I was completely terrified. I felt like a large serpent lake monster lived in the rocks under my feet and these were my last moments. And then it occurred to me, that bout of imagination pretty much summed up my 2022! Even with all the stuff life threw at us, we still had our rock. South Point Resort got us through another crazy year of adventure.

This year was extremely hard on my soul for another reason. The whole world was divided over Covid, and in Canada, the mandates and government overreach were just completely appalling. Trying to balance my own feelings, the business, the show and watching the world unravel . . . was something else. I could write a book on this alone, but that might be for later. I did learn one thing during this time. Authenticity and truth . . . are the new gold in this world. I am so grateful I moved towards a life of more self-reliance.

I finished the year on a high note. The ratings for *Haunted Gold Rush* were spectacular, so we were green-lit for Season 1 of *History's Most Haunted*! We were off to film six one-hour episodes, starting in

San Antonio, Texas. Investigating the ghosts of the Alamo was the perfect way to get fired up and re-inspired. And what a ride it was! Over the next three months, we filmed in San Antonio, Montreal, Salem, Charleston, New Orleans, and Bell Island, Newfoundland. Season 1 was going to be an unreal ride. Giddy Up!

Recent Learnings

1. When meeting your idols, know you will become a crazy person.
2. When dancing, cowboy boots and high heels can really scuff up a wooden bar top.
3. Know where all the breakers are on the property! Especially during weddings.

Standing on the Sunken Islands in the middle of Canim Lake.

CHAPTER 13

The Year of Water & "The Incident"—Part 1

July 17, 2023

So far, 2023 has been the year of water. The winter season was great. Bear was constantly off snowmobiling in the mountains on his brand-new machine with a turbo engine.

I always pretend to know what machine he's riding. All I know is we spend a lot of our profits on sleds and accessories. He's so passionate about snowmobiling. It's his actual obsession, so I don't argue much about it. He talks about it all year, and everyone knows he's a total sled-head. I'm happy he can do this for five months a year where we live.

Meanwhile, I was off filming the first season of *History's Most Haunted*. Yeah, you know, that TV show thing I was doing on the side. It took me all over North America for most of the winter, keeping my mind busy. We were both living our best lives in the off-season, and it seemed we had this life even more dialled-in.

The Canim Lake thaw was the gentlest ice-off I had ever seen—a gradual mid-April melt for the annual ice-melt competition I had been running each season. I get all our followers to guess the date and exact time the ice officially disappears from sight at South Point Resort. Trophy and all! I was trying not to take the gentle ice-off as a bad omen for things to come, but it was indeed the calm before the storm.

Bear and I were busy making spring plans. We were building a new office so we wouldn't be working out of our actual house anymore—a huge win for us. We were building fences, trimming trees, burning anything dry, and bustling through the usual spring chaos. Tanya was back for her fourth season and getting everything ready at the resort. Rick and Deb were joining us as Camp Hosts for the year. We were so excited to have additional help. They'd been guests in the past, so we knew they were fantastic people. Nory was busy planning spring events and summer markets. Everything was going along smoothly.

On April 22, I was in the Lower Mainland for an end-of-season wrap dinner with the film crew. The morning of April 23, the day I was heading home, I got an early morning text from Bear telling me that I needed to get a sump pump at Lordco on my way home. That didn't sound like a good thing, so I called and found out that our house pump wasn't working and our basement was inches away from flooding. Corine and Matt's cabin basement was already under a foot of water. He sent me a photo of their basement; all their summer kayaks and water toys were floating around.

I was so confused. *How can lake levels be THAT high? Is this even possible?* Bear explained that the water wasn't from the lake. It came from "newly formed" underground seams of water creating rivers under both of our places. Water was coming directly off the mountain—underground. He sent me a video of it flowing through and under the cabin.

What the actual hell is going on?! Scary Mountain was performing, and I was on a race to get the pumps and get home while Bear hand-dug a temporary solution using hoses, pumps, and other things. Luckily for us, Corine had accidentally brought up a flood pump (instead of a fire hose) during fire season, so we were set.

By the time I got home, Bear had most of the issues sorted. We spent the next week drying out all the stuff from Corine and Matt's basement. I'm sure the neighbours thought it was weird that all our ice-fishing tents were peppered on the lawn right *after* ice-off.

Thus began the water season.

Only a few days later, we had "The Incident."

There are two large creeks on South Canim. One on our property and one on Val's. I guess the Sallenbacks have a thing for old resorts and raging creeks. We'd had a great thaw season with consistent low temps and gradual thaws—amazing, but unusual.

On April 29, the temperature jumped to 26°C and all hell broke loose on South Canim.

Our creek had issues in the past, so the previous owners spent the time and money to put in new rock and increase the culvert size. Val's culvert was undersized, and after the 2021 wildfire season, we were both expecting some drama in the creeks. Especially with hers.

So, back to that 26°C day. Bear and I went over after dinner to check the flow of the spring run-off in her creek. The water was already starting to creep over and flood her pristine yard—the yard she had just raked and cleared. The yard that looked like a park. Val's property was an old resort with multiple cabins and outbuildings, once called The Peter Pan Resort. I always thought this was fitting. There were four boys in the family, and it was the place they came to play and enjoy the outdoors.

The water was coming over the road at this point, threatening her property. I called our Cariboo Regional District representative and warned them that they only had hours before private property was devastated. There wasn't much we could do, so we all went to bed a little anxious.

Bear and I reassessed our creek before dark. We could hear the large boulders roaring down the creek, smashing under the water and echoing up the valley. Jesus. We were about open in twelve days and it's always a nerve-wracking event when the creek has the potential to cause damage, but alas- this was heading into season five, so our emotions were a little more tamed.

At 3 AM, I bolted upright in bed. I swear I heard Bear's dad Owen (who had passed) in my head saying, "GET OVER THERE!"

I jumped out of bed, flew into my Bronco, and drove like a bat out of hell to Val's. As I passed our creek, I looked down and thought the water levels looked high, but it was dark, so I kept driving to her place. I arrived and saw a Dawson Road vehicle sitting there . . . and then I saw water coming across the road! Yikes! It was coming right toward her main driveway—and her main house. OMFG.

I looked around quickly, jumped back into the Bronco, pulled a massive burnout, and got back to our place to wake up Bear. I barrelled into the house like a banshee, turning on every light I could and yelling for Bear to wake up. Roxy went into panic mode and barked at everything. When I got to the bedroom downstairs, Bear hadn't even turned on a light. What the heck! I flicked it on and saw him sitting there, upright, his eyes huge.

I uttered his least favourite three words: "Get. The. Tractor."

It took him about thirty seconds to reset and get into action mode. I drove back in my vehicle, Roxy howling out the back window like a siren. Bear was right behind me with the John Deere. Down South Canim we went. We did absolutely everything we could to save Val's property. The sun came up while we were hauling ass, and we could see the absolute devastation of her property. The creek had brought down half the mountain, and a river was raging through it all. The amount of rock and gravel everywhere was shocking. A Grand-Canyon-sized trench now ran through her yard and into the lake, eroding everything in its wake. Val came out and was in shock. This woman prided herself on her gardens, yard, and property that were taking a thrashing from Mother Nature right before our eyes.

Bear managed to divert the water away from the driveway and her main house, and we focused on the four cabins and the gardens. I threw logs around my brother-in-law Warren's cabin to divert the water away, but I couldn't keep up with it. I was soaked and covered in mud and dirt. Val picked up rakes, shovels, and anything being washed through her property and into the lake. Things were lost.

Bear took rock from the yard to try to berm up an area around the buildings. It was madness.

At one point I saw the water divert to her main garden. I knew she had just planted the rhubarb days before. Covered in run-off, I screamed at Bear, "SAVE THE RHUBARB!" In that moment it made complete sense. We managed to divert the water from the garden with dirt and rock to save it. I'm still waiting for my rhubarb pie in thanks.

Val and I trying to turn off the breakers while water was rushing underneath cabins without real foundations was highly unnerving. At any point, those smaller wash houses could wash away—with us in them!

We did what we could . . . but the damage was done. What a mess. We all needed coffee and food after hours of insanity. Bear and I went home—only to see a brand-new scenario unfolding at our place in the light of day. Our creek had brought down just as much rock and gravel, and it ended up blowing out at the bottom and taking half our fence with it. We looked at each other and decided to go inside and have coffee and food first.

Afterward, we took stock and realized that it could have been so much worse. Our washout was a bit of blessing. Minimal damage, but it brought down so much gravel that it gave us an insane amount of new land. I think we might have manifested this South Point thing. "The Point" keeps getting bigger with every washout each year. I think we got an extra two feet of sand and gravel for the condo guests to enjoy out of this one.

The road on South Canim was closed for several days as crews figured out how to best clean and reopen it. But this was not the end of the water . . .

This summer, Beth, my previous colleague and now friend, and her husband Ian were coming again. They had already stayed in an RV, Cabin 1, and the condos, and they were super supportive of our business. Beth had retired, but Ian was still working. I will never forget the first time they came to South Point and were staying in the condos. I came to have a campfire on the beach with them the first night. I asked Ian, "So what do you do for work?" and he said he was in satellite communications. I couldn't believe it! "Like you design satellites in SPACE?!"

He quickly replied yes, and that was the conversation for the next four hours. Anything related to astronomy or space, I am all-in. Beth said it was the first time Ian had told someone what he did and they asked a follow up question. So, we have become close to the point where I'm constantly sketching ideas and sending them his way. It's on my bucket list to design something that will one day land outside of Earth's atmosphere.

This year, he brought his HAM radio and gave me a lesson. It was so fascinating. He said Canim Lake is actually one of the clearest points you can listen from. Wow! It feels so Jodie Foster in *Contact*. I'm hoping to time his next visit with a pass from the ISS and have my own call sign by then.

My previous job had landed me a ton of friends who had made their way to the resort. Tisha and her family had come up multiple times, and her son Jack proclaimed that he wants to run a resort one day. That kid is smart enough that I'm pretty sure he could do anything he put his mind to.

My friend Colleen made an appearance the same weekend we got Roxy back in 2019. She came with us on an epic trip to Bobb's Lake, where she showed us her mad fishing skills. My friends Nicole (both of them!), Jo, Joan, Christine, Alicia, Tanya, Christina, and Beth had come up to check the place out. Lynn had come up for some sunshine and paddleboarding. Jenelle came for some wake-surfing with her family. Tammy and Jasmine made their way up several times.

Colton (a fellow Elder Millennial) also left the company and was opening a brewery in Kelowna with his wife. We now swapped hilarious post-corporate self-employed stories. They had come several times to support us and experience Canim Lake. Kristy had come up for parties, family trips, and end-of-season road trips with me and Bear. It has been incredible to see the support from my old colleagues.

Destruction on Val's property from the spring run-off.

CHAPTER 14

The Year of Water—Part 2

October 5, 2023

As water seemed to be a theme this year, Bear and I were very pro-active and did some preventative maintenance around the property. We spent the money and got all new concrete lids on all the older septic tanks, new risers, and did some overhauls in the septic department. Orville's Septic spent a week here with their machines. We hoped to avoid another Shitter's Full shitstorm.

We also fixed a leak in the wall of the RV washroom and put new vanities and a new hot water tank in there to avoid any water issues in that building. We had our season opener on May 12 with all the locals, and we had a great party in the Tack Shed. Lots of music, dancing, and good times. The year was looking awesome. In addition to Tanya, we had grown our staff substantially, including welcoming Rick and Deb as Camp Hosts and Nory again in Events and Outreach. We had just been voted "Best Tourist Spot" and "Best Staycation Spot" for the second year in a row by South Cariboo residents, so we were feeling good as we headed into season five.

May long weekend was another jam-packed one, and the weather was amazing. We also kicked off our long weekend markets with vendors, food trucks, and a ton of locals showing up in droves to support us. June was drama-free. Lots of great groups came in, including tree planters, reunions, weekend warriors, and class reunions. My mom came up for two weeks, so we did lots of touring around and found our new favourite ice cream shop at the Country Pedlar at Interlakes.

Circling back to the year of water, along came the July long week-end. It was July second, and our biggest community market yet was happening in the Tack Shed later that afternoon. We had about twenty-eight vendors coming and we'd promoted it far and wide in the Cariboo. We were expecting hundreds of people to attend. Bear and I were up early because we had so much to do to ensure every-thing was ready. We had just sat down for breakfast. We remarked to each other how much better year five felt—less stressful, since we'd upgraded so many things that didn't work before.

Well, not two minutes later, at 7 AM, there was a knock at the door. Our door. We looked at each other. I opened the door and one of our guests said, "Oh hi, Leanne. I'm so sorry to bother you, but the kids said there's a flood in your bathroom."

I think I processed what she said, but I can't be sure. I assured her she was wrong and there was no flood. Maybe a small leak. I thanked her and said I would check it out, and then I shut the door calmly.

I finished doing my hair, finished drinking my coffee, and was super leisurely in my pace to go investigate this so-called flood. In fact, I was so sure this lady was wrong that I was checking the daily weather reports on my phone as I headed over.

As I walked over the bridge, I saw it. The giant streams coming out both sides of the washrooms. It was definitely a flood. *What are the chances of it happening today?!*

I looked in both doors and rivers were coming out of the walls. Seemed manageable. In seven hours, hundreds of people would need this building to go to the bathroom. Perfect.

Bear and I went to work. I also texted Ian, and he was there to help in less than ten minutes. As expected, the issue was diagnosed and fixed shortly afterward. All was good for the market.

Matt and Corine were up, and generally when they are visit-ing, Bear and I eat very well. So do the dogs. Matt is either barbe-quing or smoking some sort of thing in one of his two smokers on his patio. On a stressful day, I can usually rely on Matt or Corine saying, "Dinner is at six." I just nod in appreciation. This was one of those days.

Also this week, we had several storms and a few spot fires on Canim Lake. Helicopters were out, fire departments dispatched, and our Canim Lake Facebook Community Group was abuzz. As I sifted through some comments, I saw someone had said, "Why does the lighting always strike that mountain on South Canim?"

Someone else responded, "It's a giant iron ore deposit, and that mountain is highly conductive."

I burst out laughing. Whether or not it's true, the fact that Scary Mountain could be a legit lighting rod is hilarious to me. *That* wasn't in the resort sales brochure.

Dianne Carter and her family were back again this summer (they haven't missed a year!). One night, she and I were in Cabin 1, debriefing on *History's Most Haunted*. We were talking all things paranormal and then she looked up, pointed behind me, and said, "DID YOU SEE THAT?!"

I was like "Oh god." I didn't know what to expect. I turned around and saw a little mouse scurry across the floor of the cabin. Dianne wasn't super mobile at this point, so I was trying to corner the thing based on her very clear directions. I ran back to the house to get a trap and another broom, and then she and I were on the hunt. Eventually, I just couldn't handle the stress, so I went to get her son Mark, who was out by the campfire. We quarantined the area where the mouse came in and called off the hunt for the night.

Circling back to the year of water, *nothing* could compare to the next incident. We'd now entered South Point Resort's 2023 wedding season with four large weddings to rock out. Jamie and Ang are friends of ours who decided to get married here. They're a local couple, so we knew a lot of the guests from around 100 Mile House. The party was going to be massive. The weather looked amazing and everything was going according to plan.

This was also the night our Salem episode was airing, and I couldn't watch it because I was so hyper-focused on the success of this wedding. Our team was even on the cover of *TV Guide* that week! I was managing social media for the show, responding to comments, and sharing content. I was dialled-in on all fronts. Focused.

Decorations were up, the ceremony was about to start, and I was calm. I was in the office when, suddenly, a guest ran over from the condos. The pace of his run and the look on his face were so concerning that I quickly turned off my emotions and prepared for whatever the universe thought I deserved that day.

Then he uttered these haunting words: "You have septic overflowing and running out all over your grass in front of the condos."

I studied his face, waiting for him to smirk or laugh. He did not. I quickly pushed him out of the way, flew out the door, and saw it. I glanced at my phone for the time. *One hour before the ceremony.* Bear just happened to be over at his mom's getting something, so I made two phone calls without any hesitation. The first was to Orville's in an absolute emergency panic voice: "It's Leanne at South Point. This is one of those times. Please send the cavalry." Then I called Bear to get back here.

Some guests were in the shower getting ready for the wedding, but we decided to turn off the water because the tanks were obviously full. More water equalled more overflow. I was so grateful the bride was in the far cabin getting ready. It was in a completely different area. My goal was to keep this from them as much as I could. Ron from Orville's showed up with the massive pump truck within thirty minutes, but we made the executive decision to wait until after the ceremony to start pumping it out. I didn't need "wafts of shit during ceremony" added to my online reviews for wedding season.

The diagnosis was that both of our septic tank pumps failed. At the same time. On that day. *WTF. How do both fail? Is this the universe messing with me?!*

Regardless, this was time for serious strategic planning. I watched the wedding ceremony from the bridge . . . and I spotted a black bear on our property across the street. I didn't know what to be more worried about: the bride finding out about #shitgate2023 on her wedding day, a bear walking down to the beach during the wedding ceremony, or how the hell we were going to pump out these tanks and turn the water back on with no one smelling it. My mind raced. My heart raced.

I breathed a (non-shitty smell) sigh of relief at the end of the ceremony. Because really, that was what they came here to do. They were married now.

I basically bribed the photographer to take the wedding party across the street and keep them there for a while. During this time, we assessed which way the winds were blowing and started pumping. It took about forty minutes to deal with the situation. I was so thankful to those guys for the quick work. They had to come back the next day and install new parts as we needed new pumps. Basically, any profit from that wedding went toward the repair bill for this one—and then some.

Jamie and Ang were so good about it, and I was grateful for that, too. It turned into a beautiful wedding, and the party that night was just epic.

The last wedding of the year belonged to Madi and Joe, two of the best humans you could ask for. We had already become close with the bride's parents (Peter and Ang), as they had already camped with us twice that year. Peter used to work at the Vancouver Fire Hall with Bear's dad back in the day, so it was a cool connection.

It was the last weekend of September, so the weather can always be a bit touch and go, but you hope for the best. The night before the wedding was gorgeous on the beach and the rehearsals were fantastic.

The morning of the wedding, the bridal party went into town for hair and makeup, and it was left to the boys and a small team to get everything else ready. I used to work with Madi's aunt Tammy at my previous job, so we were reminiscing, decorating, and having a great time.

Suddenly, I got a weather alert on my phone. *Intense rain* was forecasted for the afternoon. *Uh oh*. It was right at the time of the ceremony. We held off putting out the chairs for as long as we could, but eventually we needed to set everything up.

I'll never forget the moment it happened. I was in my house, boiling the kettle, when I heard it. Monsoon rain, sheet rain, or whatever you want to call it, had commenced. I heard it bouncing off my metal roof in the house and I was horrified.

I sent Tammy a text: "It's okay. I'll bring towels."

As I glanced out the kitchen window, I saw the bride, her mom, and her bridesmaids arriving back from town in their car. *OH. MY. GOD*. I grabbed my jacket and an umbrella, and I ran in my big red gumboots toward their vehicle, parked by the cabin. No one was getting out. I offered them my umbrella as they opened their doors.

Madi's mom yelled, "What the hell is happening?!"

The bride was focused on getting inside the cabin without messing up her hair and makeup. Once they were all inside, I ran to the Tack Shed where I'd left Tammy. As I got there, I wanted to laugh, but I just couldn't. The rain was so intense that it penetrated the *metal roof* of the building. It was pouring all over the just-decorated tables. I had no idea how it was getting in, but I foresaw a new roof for 2024.

Tammy had improvised, covering up the tables with the paint-class tablecloths she found under the bar. I threw towels all over the floor, trying to soak up the water, and turned on all the space heaters.

There was a moment when the two of us were just standing inside, looking down at where the ceremony was supposed to be in minutes, watching the flowers on the arch being pelted with rain. She said, "Someone better tip that florist. That is amazing work staying on that arch."

The rain subsided. We were soaked. The floor was soaked. And the tables were in an unclear condition. I ran down and dried all the chairs for the ceremony and anything with a drop of water on it. Then I went back to the Tack Shed and started peeling painting cloths off the tables and assessing the damage. It wasn't bad. Tammy had really saved it. I brought more towels, cleaned the floors, and made sure nothing was damp. I think I burned 3000 calories in about twenty minutes.

The wedding was amazing, and the reception was one for the books. It's vivid in my mind because the people were so fantastic, and they really let me be a part of it. It was such a wonderful way to close the 2023 season and wrap up our year.

The day after the guests from this wedding checked out, I headed to the Coast to fly out to New York for the US launch of *History's Most Haunted*—a small press event in Manhattan and a paranormal convention in Plymouth. Resort-brain off, ghost-hunter brain activated. This was my life—the constant juggle between the wild existence of running a resort and doing interviews with major press outlets all over the world for ghost hunting.

Recent Learnings

1. Always have a bin of towels by the door, ready to go.
2. Believe your guests when they say there's a flood.
3. Make sure you and your weather app are best friends on wedding days. And pay the extra to track the radar.
4. Pumping out septic on a hot day can sometimes smell like Thai food (according to guests).
5. Make sure you're on a first-name basis with your septic company and always pay the emergency fee. Don't ask, just pay.
6. Wear running shoes or rubber boots on the day of a wedding.

Ryker, Roxy, and I at the sand dunes at Farwell Canyon.

CHAPTER 15

Reflecting on Five Years

February 2024

We had an awesome fall season quadding in the woods, exploring the mountains with friends, and enjoying what the Cariboo has to offer. We love exploring and finding things off the beaten path. We look for those gem spots that are just too good to post on social media for fear of them becoming popular.

One afternoon, we were out with Bob and Natalie near Spanish Mountain, intending to hike into the old Flour Mill Volcanos. We had packed a lunch and were determined to find this old site. I was driving us down a narrow trail in my Bronco, when suddenly something was on the road in front of us, hauling ass. We all thought it was a moose, but it was actually a wolf! We were going forty-four kilometres an hour, and this thing matched our speed and was almost running alongside us. Amazing! You have to go into these unknown areas to have those experiences.

The wolf darted off the trail and into the dense forest in front of us, and then Bear yelled, "STOP! I think this is our trailhead." I wondered if the wolf was just luring us to his pack. We took the risk anyway, bushwhacking through the forest to find the place we'd been looking for.

The wolf is symbolic for me and our lives here. This entire adventure has been about taking a risk and going into the unknown, following paths not knowing where they lead, and trying to outrun the world and survive these last several years. The day we all spotted the wolf solidified our group as "The Wolf Pack."

The endless renovations, repair bills, system failures, and stress don't even matter. We have had way more laughs, met new friends, helped build a community, and improved our lifestyle one hundred percent. Waking up and looking at the lake every day is paradise to us, and being able to make a living to help us do that is worth it. I often think of what I would be doing in the office at a certain moment, or if I'd be sitting in traffic on my way home from work, letting the stresses of the rat race consume me.

Bear and I enjoying a sunset campfire.

This journey isn't for everyone, that's for sure. However, if you find yourself off-brand or living a life that doesn't fulfil you, I highly recommend making the leap. Maybe not into resort management, but into a life that makes you see the stars a little clearer.

With drought looming in 2024, a lower-than-normal snowpack in the hills, and forecasts of an intense fire season, I'm a little worried. Insurance costs for resorts are out of control, property taxes are on the rise, and resort operating costs have more than doubled since we took over. However, I know there's nothing we can't face at this point. And hell, even if we lost it all, we would just move on to the next big adventure together. We're investing in our own fire suppression equipment, and we're building South Point Resort into something even more spectacular—a place people will hopefully enjoy for years to come.

At one point, Canim Lake had over twenty-five resort properties on it. In the 1950s, it was a trophy fishing lake with tons of celebrities coming to stay. The likes of John Wayne, Marilyn Monroe, President Hoover, the Rat Pack, and others graced these waters. There is so much history here, yet today only four resorts remain. South Point Resort, Rainbow Resort, Reynolds Resort, and Canim Lake Resort are the final pillars of the Cariboo resort rush on Canim Lake.

The year 2024 marks seventy years this property has run as a resort. Who knows what the future will bring? But I would like to contribute to a little piece of history here on South Canim Lake. Even if it's just a living memory for a young couple, sick of the city, attempting to carve out a new life. I'm curious to see how the lake looks in another twenty years. It's cool to see our nieces Isabella, Scarlett, Maddison and Evie enjoying the lake. I know Canim will always be a big part of their lives.

There's still lots of work going on in our off-season this year. We just wrapped up renovating Cabin 2, painting the last of the interior condo walls, clearing a new trail access across the street, and burning endless brush piles. We've taken advantage of the lack of snow to rake the entire property, cleaning up a bunch of brush to avoid lots of work in the spring. Ian bought a new excavator, so I'm sure there are more projects on the horizon.

In early January, Jen, Tanya, and I spent ten days in Waikiki, soaking up the surf, sand, and sunshine. We talked a lot about goal setting and our big plans for 2024. Tanya has told us that this will be her last year with us, so we're looking forward to rocking out one final amazing season with her and seeing what adventure she's off to next. That will be a big loss for the resort, but we will welcome new staff in the coming years and the turning of the page.

I moved up here when I was thirty-four, and I'm turning forty this year. Remember when you were young, and you thought forty was ancient? I can't believe it. I'm definitely going to embrace the next decade and whatever it brings. Going into year six at the resort, we have some big questions on our mind. We have to decide on expansion plans or settle in with what we have. The business and economic landscape in BC is ever-changing, so planning anything in the future is pointless at this stage. Turning forty, however, it seems like some life reflections and personal goals are in order. I hope *History's Most Haunted* continues, and I hope that the next five years are just as entertaining as the last.

I've talked to a few other younger couples who have reached out and asked for advice on resort management. They want to do the exact same thing: leave the rat race, their good jobs, and jump into a change of pace out in nature. I tried to be realistic. It's a lot of work and maintenance, but overall, their lives would improve and they would be happy. They all ended up purchasing properties, and I hope they don't regret it.

I find it interesting that I have become a "purchasing a campground" consultant for people. I even had a couple from the USA reach out and ask me questions about the pros and cons of doing this. It might be a global trend. People aren't tied to the archaic regime of the prescribed life anymore. They're willing to do something a little different for themselves. Create something for future generations to enjoy by reporting to themselves and building their own dreams, not someone else's.

Remember when, as a kid, they asked you what you wanted to do when you grew up, and we all said something really cool? Why does becoming an astronaut suddenly seem out of the question when you enter high school? I mean, I was born for Starfleet, so I'm hoping that, in the next life, I can really excel at that career choice.

As I wrap up this chapter and this story, I see Bear outside on the lake, auguring a hole for some ice fishing. It's a beautiful sunny day in the Cariboo, and burbot is on the menu for dinner. Whatever the next five years bring South Point Resort, know that, so far, we have had an absolute blast. Thanks to everyone that was a part of our beginning, and we look forward to meeting everyone who will be involved in the next chapter.

And remember . . . hug your dogs, take the risks, watch the skies, and wildly follow paths into the unknown.

Leanne

A shot of me under the northern lights in front of our house.

Acknowledgements

I'm indebted to my editors, Tara Avery and Kelly Ireland, for their invaluable insight and guidance. I am thankful to Carol Watterson at Cubehouse Productions and Jan Westendorp at Kato Design & Photo for their expertise and support in bringing this book to life.

We are so grateful for every single guest, neighbour, friend, family member and anyone who was a part of these first five years at South Point Resort. It wasn't what we expected, but your presence, encouragement, and belief in us made it possible—and gave us the confidence needed to keep going.

Bear—thank you for this incredible life we share. There is no one I would have rather navigated this wild journey with. Thank you for introducing me to Canim Lake and co-piloting this crazy ride every step of the way.

About the Author

LEANNE SALLENBACK has been an outlier her entire life. Described as industrious and determined, she's always pushed boundaries, asked questions, and constantly examined the status quo. With a BBA in Marketing Management, she spent more than 12 years working in the tourism and energy sectors before quitting corporate life, selling everything, and moving to British Columbia's Cariboo region to pursue a lifestyle in self-reliance. If revitalizing a rustic, 70-year old resort and transforming it into a multiple award-winning destination wasn't enough, Leanne's also found time to star as a co-host and paranormal investigator on T+E's *History's Most Haunted* TV series. In her spare time, Leanne loves travelling, ancient archaeology, the supernatural, spending time outdoors with her husband and dogs, and gazing at the night sky.